Lecture Notes in Mathematics

Edited by A. Dold and B. Eckmann

T0215391

450

Algebra and Logic

Papers from the 1974 Summer Research Institute
of the Australian Mathematical Society,
Monash University, Australia

Edited by J. N. Crossley

Springer-Verlag
Berlin · Heidelberg · New York 1975

Prof. John Newsome Crossley
Department of Mathematics
Monash University
Clayton Victoria 3168
Australia

Library of Congress Cataloging in Publication Data

Main entry under title:

Algebra and logic.

(Lecture notes in mathematics ; 450)
Bibliography: p.
Includes index.
1. Logic, Symbolic and mathematical--Congresses.
2. Algebra--Congresses. 3. Groups, Theory of--
Congresses. 4. Commutative rings--Congresses.
I. Crossley, John N. II. Australian Mathematical
Society. III. Series: Lecture notes in mathematics
(Berlin) ; 450.
QA3.L28 no.450 [QA9.A1] 510'.8s [511'.3] 75-9903
ISBN 0-387-07152-0

AMS Subject Classifications (1970): 00-02, 00A10, 01A05, 01A60, 02-03, 02A05, 02C10, 02C20, 02E99, 02F99, 02K05, 13B25, 13F20, 13J05, 13L05, 20E40, 20M10, 55E05

ISBN 3-540-07152-0 Springer-Verlag Berlin · Heidelberg · New York
ISBN 0-387-07152-0 Springer-Verlag New York · Heidelberg · Berlin

Offsetdruck: Julius Beltz, Hemsbach/Bergstr.

PREFACE

This volume comprises some of the papers presented at the Fourteenth Summer Research Institute of the Australian Mathematical Society held at Monash University, Clayton, Victoria, Australia from 6 January to 16 February, 1974. A detailed report follows but it is a pleasure to acknowledge the generous financial support of Monash University, The Sunshine Foundation, A.V. Jennings Industries (Australia), CSR, ICI, BHP, General Motors-Holden and the Australian Mathematical Society. Finally our thanks to Anne-Marie Vandenberg who did all the typing.

J N Crossley

Melbourne, 15 December, 1974

CONTENTS

REPORT

on the

FOURTEENTH SUMMER RESEARCH INSTITUTE

of the

AUSTRALIAN MATHEMATICAL SOCIETY

Monash University

6 January - 16 February 1974

The Fourteenth Summer Research Institute of the Australian Mathematical Society was held at Monash University in conjunction with a Summer School in Mathematical Logic (6-26 January) and a Symposium on Recursive Model Theory (6-19 January). These last two meetings were co-sponsored by the Association for Symbolic Logic. The organizing body included : Professor J N Crossley (Director), Dr H Lausch (Treasurer), Dr C J Ash (Secretary, with the (massive) assistance of Mr W F Gross and Mrs E A Sonenberg). Secretaries to the Conference were Ms C Eagle and Ms A-M Vandenberg. There were four major academic divisions : Pure Mathematics and Logic (organized by J N Crossley), Applied Mathematics (Dr R K Smith) and Statistics (Professor J S Maritz).

The Director is happy to report that having invited lots of overseas visitors, almost all of them participated in the Conference. There were a large number of invited addresses and seminars. The Summer School on Logic had 7 formal courses of six lectures each (including 3 advanced and 3 introductory courses). Running the Summer School and Symposium was obviously acceptable to a large number of people and might well be considered by future directors.

There were relatively few foundations and bodies in industry which gave support, but those which did were generous.

Donations were received from

The Sunshine Foundation	$1,000
A V Jennings Industries (Aust.)	500
CSR Ltd	200
ICI Australia Ltd	100
BHP Australia Ltd	250
General Motors-Holden	200
Total	$2,250

The Sunshine Foundation funds were given on condition that we matched the $1,000 from other sources - which we did.

As usual the Australian Mathematical Society gave us a grant ($500) which was very useful in the early stages.

Monash University generously gave $5,000 towards the Summer School and Symposium.

Thus we had a relatively large budget. Added to this we had by far the largest number of registered participants for any Summer Research Institute (253). We believe this large number was attracted by the large and distinguished collection of visitors, particularly from overseas (more than 20). Many of the United States' visitors were funded to some extent from United States sources, others were on short-term visiting appointments, principally to Monash. We are grateful to the Mathematics Department at Monash for making these posts available from its limited quota for visitors.

Mannix College was the venue of the Conference, just across the road from Monash, and proved an excellent choice. We record our thanks to the Master, the Business Manager (Leo de Jarlais) and their staff for the tremendous pains they took to see that everyone's needs were met.

At the General Meeting the Director was asked to try to arrange the next Summer Research Institute but one (1976) to be in Adelaide.

Finally, as Director, I should like to comment that the Fourteenth Summer Research Institute was far from a "quiet time for research". It had a full and, I believe, highly stimulating programme part of which was a "crash course", but I feel confident that this approach does more for mathematics in Australia at this time than a quiet period of contemplation and reflection would do. However, I think it would be unjustifiable (and quite possibly unprofitable) to have a Summer School every year (however varied the subjects). What form or forms the Summer Research Institute should take is, I believe, a very difficult question; perhaps we should not try to pin it down. But it is certainly valuable to have an annual activity involving lots of mathematicians over an extended period (though I would prefer 3 weeks to 6).

It is not possible to thank enough all those very many people who helped at the Summer Research Institute but it was a great delight to have the help of such an enthusiastic and hard-working crowd of people dedicated to the furtherance of mathematics.

J N Crossley

REMINISCENCES OF LOGICIANS

reported by

J N Crossley

On Tuesday 15 January 1974, the following met in my office to talk about the rise of mathematical logic: C-C. Chang, John Crossley, Jerry Keisler, Steve Kleene, Mike and Vivienne Morley, Andrzej Mostowski, Anil Nerode and Gerald Sacks. The discussion went as follows†.

Crossley What did you do, Steve, when you first started logic, you didn't have books, did you?

Kleene Didn't have books?

Sacks Well, he had Principia (Whitehead and Russell [1910, 12, 13]) (laughter). Let's see, was there a book by Lewis (Lewis and Langford [1932]) on model theory?

Kleene Well, I never read Principia; of course I thumbed it a little bit. Rosser, I guess, started in logic that way, but I learned logic by learning Church's system, which was subsequently proved inconsistent. Out of this system we abstracted λ-definability. It was only after I got my degree that I really began to read much of the literature. Hilbert-Ackermann (Hilbert and Ackermann

---

†We are grateful to the participants who allowed the taping and assisted in the editing of the transcript.

[1928]) was round, and the first volume of Hilbert-Bernays (Hilbert and Bernays [1934, 39]) appeared in 1934. I never read Lewis and Langford. The first we knew of Gödel's paper was one time the mathematics colloquium speaker was von Neumann, and of course von Neumann had lots of things of his own to talk about. But instead of that we got in there and found out he was telling us about Gödel's 1931 results. This was at Princeton and it was in the fall of 1931, and whether he had the paper itself or not I do not know, because there was a little conference in Germany, which I think is reported in "Ergebnisse" somewhere, with von Neumann and Carnap and Gödel and two or three other people, Heyting, I think. (Gödel [1931-1932]). It's a sort of summary or forecast of what was to be in Gödel's main paper. (Gödel [1931]) So it is clear that von Neumann was present at this thing where this was discussed.

Chang Was this the first you had heard of Gödel, too?

Kleene This was the first – when we went into this meeting. Church was teaching a logic course, Rosser and I were among the students – as far as I know it was the first that any of us heard of Gödel. I don't know whether Church was aware of Gödel's 1930 paper (Gödel [1930]) on completeness. He never gave it in class. I never had the classical form of the propositional and predicate calculi in my course work; I learned them for myself afterwards.

Sacks Was this in Fine Hall?

Kleene This was in Fine Hall. So as soon as we heard the lecture, the paper was available, so we went and we read the paper right off. Church was convinced there were sufficient differences in the way logic was formulated in his system that it would escape the theorem that you could not prove its consistency in the system itself. Of course he was right! Though it took a couple of years to find out. Well,at that point we did not have the more general proof (discovered later) of Gödel's theorem, which was connected with the whole idea of computability and the effectiveness with which you establish a formal system-that the notion of provability has to correspond to recursive enumerability and that there are sets that are not recursively enumerable. This approach to it was not available then, so you could speculate from hearing von Neumann's presentation and from reading Gödel's own paper. There was a question: "How general is this thing?"

Crossley What was the genesis of the interest in decidability or computability at that time? Were there informal concepts floating around that people tried to formalize or what?

Kleene I don't know how it looked to Gödel, because I did not meet Gödel till the spring of 1934, when he gave us a series of lectures (at which Rosser and I were the note takers (Gödel [1934]). In this series he brought out the general recursive functions, for which he said some of his motivation had come from Herbrand. So Gödel had discussed these things with Herbrand. How it looked to him and how long that had gone on (maybe over two years) I do not know, but I know how λ-definability arose. I took my first logic course in the fall of 1931, and I was the note-taker in that logic course. I was busy taking the notes, although I would say some things. Mainly I was just assimilating the material. The moment I got done with the course, I wanted to work on something, and Church had introduced his system and he had introduced the definition of the integers in it. You all know λ-definability here. So the question was to develop the theory of natural numbers or positive integers on the basis of his postulates. Well, actually to develop it on the basis of a subset of his postulates. He had the description operator and I wanted to develop it without that. So I had to develop the Peano postulates. How was I going to develop these? One of them, I think it is number three- anyway it is the one that says that, if the successors of two numbers are equal, the numbers are equal- and if you have a predecessor function, the proof comes right off. So I set out to get the predecessor function, and the first time I tried it, I suppose I did it in a couple of hours. I could not do it with his system of integers, so I got a different system of formulas for the integers, and with that I got a predecessor function. Well, it turned out that — (you know I was a pretty callow graduate student at that time) —, as far as proving Peano's postulates that would do fine, but that did not take care of recursive definitions which his system of integers was designed to facilitate.

So I had to do it with his system. So I went to the dentist one day and he pulled two wisdom teeth, and while I was in the dentist's office I figured out how to do the predecessor function. It took me, I suppose, a week to finish the proof in the system that it had the properties I wanted it to have. I took it to Church, and Church said that he had just about convinced himself there wasn't a predecessor function. So the initial idea of how much we could do in λ-definability was so limited that we didn't think we could do the predecessor function $x \dot{-} 1$. So there was no idea in the beginning that this was going to be all effectively calculable functions. But I kept taking it as a challenge and everything I tried I could work, and then of course we got general methods, and so it was an unexpected fallout from Church's functional abstraction operator and the definition of the integers from it that we got the idea that this could represent all calculable functions.

Nerode When was all that finished?

Kleene The basic work was done between January 1932 and probably the next five or six months. In the next year I was writing it all up. I had everything I had tried. I mean every kind of function I had tried to define, and every kind of effective operation that I had tried to parallel by λ-definability, I probably knocked off within the first five months. I think it was maybe a year later that I got the recursion theorem in terms of λ-definability. It probably was a full year later, and then Gödel arrived. I was away from Princeton in the fall of the academic year

1933-34. I am not sure whether Gödel arrived just for the spring semester or whether he had already been there in the fall; but anyway I was away from there in the fall, and I came back in the spring and Gödel was giving lectures. Gödel had this notion of general recursive function, and the questions came up:

"Does this embrace all effectively calculable functions, and is it equivalent to λ-definability?"

As I said, for us the first idea that λ-definability was general was after the fact — after having formulated that the λ-definable functions are simply the ones for which you can find formulas in this symbolism, and discovering that everything you thought of that you wanted to prove λ-definable, you could. It was Church (I have to give the credit to Church, I can't take it myself) who asked whether we had not really got <u>all</u> the effectively calculable functions. Then Gödel arrived on the scene with another concept, and there must have been discussions between Church and Gödel. I do not know how ready Gödel was to embrace the thesis that they were all the effectively calculable functions. But Church, of course, was the one who certainly came out explicitly with this, and then it was a simple matter to prove the equivalence of the two notions. And then we had done all this work before we heard of Turing. Turing's paper [36-37] is also of 1936, like Church's one [36] with his thesis, but a little later in 1936; and my impression is that Turing did it independently of knowing anything about what we were doing in Princeton. We certainly did our work before we heard of what Turing did. Post had another version, which he did round about the same time; I guess Post probably knew of what we were doing in Princeton.

Keisler Was Church's thesis just a sort of off-hand remark, then?

Kleene He spent some months sweating over it and saying: "Don't you think it is so?" and I was a sceptic, and when he came out and asserted the thesis I said to myself: "He can't be right." So I went home and I thought I would diagonalize myself out of the class of the λ-definable functions and get another effectively calculable function that was not λ-definable. Just in one night I realised you could not do that, and from that point on I was a convert. But until Church really came out and said so, I guess I had not really believed they would be all of them.

Sacks Didn't you have some feeling of getting the recursion theorem too?

Kleene I think I got the recursion theorem just a little bit later than that. I got the recursion theorem before I left Princeton in June of 1935, and of course we already had Church's thesis in the late spring of '34 – that is when Church was talking with Gödel about his general recursive functions.

Mostowski Do you know the way in which Gödel (I can't remember exactly which year) introduced the notion of a function in an axiomatic system of arithmetic and where he says that unlike the other notions it is independent of the system?

Kleene (to JNC) Yes. You don't happen to have a copy of "Introduction to Metamathematics" (Kleene [52]), do you?

[A copy was produced.]

Kleene The reason is, I cannot remember what year it is, but I think I could spot it...

Morley Yes! That is in that book; I have read it many, many times.

Mostowski I think it was earlier than what you call Herbrand-Gödel. I don't know of Herbrand making this...

Kleene What Herbrand did in the general recursive functions, as presented by Gödel giving credit for ideas of Herbrand, is something (I understand) more than what Herbrand published. What he published was a little note, or just a short piece at the end of something else. Of course, he says it is independent of the system, and whether that is saying it is the same as the intuitive notion of effective decidability ...

Mostowski No, that he certainly does not say. That is another thing.

Kleene I know just where to find this thing. I think it is what I call *resolvable* in a system (*resolvable* predicate, p.295 of Kleene [52]).

Sacks It is hard to remember everything in that book, isn't it?

Morley I did it for a week or two when I had to take my exams, but I must say a few facts have slipped my mind since then.

Kleene It helps to have written it. You can remember what is in it.

Sacks You know Rosser made me go through it from beginning to end.

Chang I don't even remember what is in "Model Theory" (Chang and Keisler [73]) anymore.

Vivienne M. That is because there are two authors and neither of them have to assume responsibility.

Kleene I thought I would get a reference here to Gödel having claimed that resolvability in this particular system was independent of the system. (Pause)...

Let me see ... "Remarks contributed to a discussion "Diskussion zur Grundlegung der Mathematik", vol. 2 (Gödel [31-2]). That comes from that conference where Gödel, von Neumann, Carnap, Heyting were present, which is a little earlier than Gödel's [30]. Actually, the printing of [31-2] is a little bit later, but the conference was a little earlier. I would have to look it up to see which article it was in.

Mostowski I think that is in this paper "Über die Länge von Beweisen" (Gödel [36]).

Kleene "Über die Länge von Beweisen", which is 1936.

Mostowski Yes, that is with this note added in print. That is this note. How does ...

Kleene As I say, I do not know how firmly convinced Gödel was that his general recursive functions represented all effectively calculable functions.

Sacks You seem very sceptical.

Kleene I think he was sceptical, and it may well be that Turing's presentation also brought Gödel around.

Nerode Actually he said that in print somewhere, I mean, I read that in the last year or two.

Kleene Ah, let me see, on some of these things of Gödel like the notes that Rosser and I took in 1934 (Gödel [34]) and the 1948 Princeton Bicentennial, for which Gödel generated a paper (Gödel [46]) (which was to be published, but whoever was to edit the thing never did get it published, and I made a copy available to Martin Davis for "The Undecidable" (Davis [65])). I believe in Martin Davis' volume there are some notes added by Gödel, which I do not know that I ever got around to reading, but that could well be in there.

[A copy of "The Undecidable" was produced.]

The Kleene-Rosser notes on Godel begin on p.39. On p.71 they end and we find a "Postscriptum. In consequence of later advances" – this was contributed by Gödel, whenever it was necessary (early 1960's) for Martin Davis. "Postscriptum. In consequence of later advances , in particular of the fact that, due to A.M. Turing's work, a precise and unquestionably adequate definition of the general concept of formal system can now be given, the existence of undecidable arithmetical propositions and the non-demonstrability of the consistency of a system in the same system can now be proved rigorously for *every* consistent formal system containing a certain amount of finitary number theory.

"Turing's work gives an analysis of the concept of 'mechanical procedure' (alias 'algorithm' or 'computation procedure' or 'finite combinatorial procedure'). This concept is shown to be equivalent to that of a 'Turing machine'," and so forth. Yes, so I think it was Turing who overcame Gödel's doubts on the generality of these concepts. Except of course in that 1936 paper "Über die Länge von Beweisen", Gödel certainly does say that the notion of what functions you can represent is independent of the system. But that is not quite the same thing as saying it is all effectively calculable functions.

Mostowski — He already knew the notion of the so-called Gödel-Herbrand computable functions.

Kleene — Yes, but whether he agreed that it was completely general from the point of view of algorithms or finite procedures or that, I guess ...

Sacks — He is a very cautious man.

Crossley (to Mostowski) — Were you involved in recursive functions at that time?

Mostowski No, no, I learnt.

Sacks He was just a little baby then.

Mostowski Not quite so. I learnt them from Steve's paper which I possibly had learnt of in Mathematische Annalen (Kleene [36]), so the theory was already quite well developed.

Crossley Were you a bit later than Steve in getting your degree?

Mostowski Yes, I studied this paper in – when was it published?

Kleene Published in 1936.

Mostowski So I studied it in 1937.

Crossley Did you do a thesis?

Mostowski My thesis was on the axiom of choice.

Kleene With whom?

Mostowski With Tarski.

Chang Yes, so-called Fraenkel-Mostowski models.

Mostowski So-called Fraenkel-Mostowski models ...

Nerode So-called.

Sacks Could you explain that joke?

Keisler So when did you finish your thesis?

Mostowski I finished it early in 1933. Then I went to Zürich.

[There is a gap in the tape here of approximately 18 minutes.]

Mostowski Some more early history. Who started this business with set theory and recursion theory? I mean, as you pointed out in your lectures this, what people now call recursion theory has nothing to do with recursion theory and I quite agree with that. But who got the idea that these set-theoretical operations are recursive operations?

Sacks I am no expert on the history, but ...

Mostowski You made it!

Sacks I think Takeuti started it – you mean set theoretic operations as distinguished from recursion theory on ordinals?

Mostowski Set theory and ordinals I think that is quite obvious that it is a generalization. I don't know whether Machover (Machover [61]) was earlier or Takeuti (Takeuti [60]).

Sacks No, Takeuti was earlier.

Mostowski But that is obvious. But then came this fashion that one speaks of L and the operations in L as a generalization of recursive operations.

Nerode That is the part that Takeuti did; the business on L , and it was not until at Kreisel's suggestion (I think at Kreisel's suggestion) that both Kripke and Platek thought of restricting it to initial segments that the other part came about.

Sacks Well, in the last two pages of Takeuti's paper he does consider countable initial segments of the ordinals.

Mostowski Of the ordinals, that is all right. I think that is still this obvious generalization but I think the very, very essential generalization was from ordinals to sets. What is recursive in the operation of forming the union of sets?

Sacks Yes, there is a mysterious footnote in his paper. (It really is early, by the way, the date of receipt on it is 1959. It is definitely earlier than Machover, Levy and all the rest.) But there is a mysterious remark in it on the first page in which he thanks Professor Gödel for some important insights – he does not say what they were, and then a lot of the notation in that paper is the same as the notation in Gödel's continuum monograph.

Nerode On the other hand, the fashion of doing it in terms of L certainly started with Dick Platek's thesis. In other words, when he wrote the thing on admissible sets that was done entirely in the set-theoretic language for the first time because Kripke did it in the other language.

Sacks No, Kripke stuck to L.

Nerode Kripke stuck really to equations on the ordinals whereas Dick Platek's was entirely in terms of fragments of set theory.

Sacks Right.

Nerode And Kreisel, I think, also originally had it entirely in terms of ordinals and notations for hyperarithmetic sets of ordinals, so I think the first place where it is absolutely explicit is the Platek dissertation. I am not absolutely sure.

Sacks I have never understood why Barwise calls it Kripke-Platek – calls it KPU meaning "Kripke-Platek with urelements" – Kripke gave a course at Harvard, I mean, I have a copy of the notes taken by a student. Now that course was given after Platek got his Ph.D. In fact that year Platek was an instructor at M.I.T., so there is no question about it.

Crossley Oh, but that was in 1964.

Nerode Just wait till the conclusion.

Sacks Now the point is ...

Nerode He does not have it in there.

Sacks He has something like that in there, but he makes the mistake of using replacement rather than, well, Platek called it reflection, in my lecture I called it bounding. In other words, he makes the mistake that you would make if you were thinking in terms of L where there is no difference. So I think Platek really was the first to put together the notion of admissible set.

Nerode Exactly what Mostowski is talking about is what Platek did, I think, for the first time in his thesis – the business of not just having recursion equations on the ordinals or having primitive recursive functions on the ordinals, but actually saying the set operations with restricted replacement and reflection.

Sacks: Takeuti took a strange route. He has a paper in the early 1950's somewhere (Takeuti [55]) in which he has a system of set theory in which everything is ordinals - in other words, instead of talking about sets you just talk about ordinals and he has some theorem to the effect that in some sense his system is equivalent to ordinary set theory, and then his paper "Recursion Theory on Ordinals" is really a sequel to this, at least in his mind.

Keisler: Who would you credit the notion of an admissible ordinal to?

Sacks: Kripke certainly had that, I mean, he may even have had it before Platek. Maybe it was independent.

Nerode: But the notion of admissible set is not Kripke - Platek, that is Platek.

Sacks: That is really Platek, because Kripke very definitely made this mistake and it is a very simple mistake to make, insisting on replacement rather than reflection.

Nerode: But remember when that contradiction turned up for a week or two - when you were still at Cornell?

Sacks: Graeme Driscoll was my very first student in this area, recursion theory on the ordinals. I thought I would start him off by having him think about Post's problem. First of all we confined ourselves just to the recursive ordinals, in other words, the very first admissible ordinal, and Post's problem had already been done there, so I thought he could start off by reflecting on it, perhaps improving it a bit, getting a tidier solution, and he did get a very efficient solution, much more so than the existing one, and he kept thinking of further improvements day by day until finally he came in with a contradiction. You see, initially he built two sets which were incomparable. Then he thought of ways of giving them additional properties. If they are incomparable and have property P then they have property Q as well, but they are still incomparable, and then, the final property was comparability. He could make them comparable.

He retained the original construction for making them incomparable, but then he had another construction – as each one developed, you sort of poured it into the other – and they would be comparable. I could not find any mistake! That was a lot of fun. And I called up Kripke, and his reaction was that there was this contradiction in the subject and that the whole thing was going to collapse.

Nerode But it was simply due to different definitions of reducibility, which looked as if they would coincide, but because they do in the finite case, no-one had thought it through.

Sacks In fact that mistake led to a whole bunch of interesting things. In fact, I think the last theorem in that series was just a few months ago, due to Richard Shore – one of the final in the series. He sort of explained what goes on for all admissible ordinals as far as this phenomenon is concerned.

Mostowski So why don't you publish Platek's Ph.D. thesis?

Sacks There is no need to publish the part on admissible sets, I think. But it would be interesting to publish the part on higher types. He had certain proposals and results about recursion in higher type operations.

Kleene Has he published later things?

Nerode No, he has written later things and left them in the same unpublished state.

Sacks Oh, he did publish one paper on the super-jump (Platek [71]).

Nerode Which was wrong. It has given rise to a great deal of work because he made a mistake in it. In fact, I think all the work ...

Kleene Valuable work, maybe?

Nerode I think the later work is very valuable. Actually one mistake, which no-one that brilliant could normally make. He made a mistake and has given rise to a great deal of work. The Aczel-Hinman work (Aczel-Hinman [74]), and your student, Leo Harrington have all picked up on that.

Sacks And Harrington's result.

Mostowski I must raise a protest against this habit of not publishing. That is all right here, because you meet every second month to collaborate at a conference here or there, but people like me are completely cut off.

Kleene Yes.

Sacks I agree completely.

Morley It should lapse.

Sacks And anyone who does not publish his work should be penalized.

Kleene This is just what I wrote to Emil Post, on construction of incomparable degrees and things like that, and he made some remarks and hinted at having some results and I said (in substance): "Well, when you leave it this way, you say you have these results, you don't publish them. The fact that you have them prevents anyone else who has heard of them from doing anything on it." So he said (in substance): "You have sort of pricked my conscience and I shall write something out", and he wrote some things out, in a very disorganized form, and he suggested that I give them to a graduate student to turn into a paper. As I recall, I think I did try them on a graduate student, and the graduate student did not succeed in turning them into a paper, and then I got interested in them myself, and the result was eventually the Post-Kleene paper (Kleene and Post [54]).

Morley You mean one of your graduate students could have had the Post-X paper?

Nerode You mean one that wanted to work hard.

Kleene I suppose, maybe it was not good for a graduate student, because a graduate student needs a thesis he can publish under his own name, and this would have had to be joint, or maybe ...

Morley — Oh, I don't know. I think he could have borne having a Post-X paper.

Kleene — As a matter of fact, it could have – if a graduate student had picked it up. There were things that Post did not know, like that there was no least upper bound. You see, Post did not know whether it was an upper semi-lattice or a lattice. I was the one who settled that thing.

Sacks — What are you talking about? The degrees of the arithmetic sets?

Kleene — No. The upper semi-lattice of degrees of unsolvability. So if a student had done those things, he could have put his part into his own paper and another one into ...

Crossley — How did Post come in, anyhow? Whose student was he?

Kleene — He took his degree about 1920.

Nerode — With Sheffer?

Kleene — H.M. Sheffer.

Mostowski — Look, Tarski told me the following story about Post, which I believe is true. After arriving in the United States, he told Post that he congratulated him because he, Post, is the only logician who has made important contributions to propositional calculus and who has nothing in common with Poland. Post's answer was : "Oh, no, I was born in Białystock, and that is a town in the east of Poland."

Nerode That is a new fact; Post was a Polish logician.

Kleene When did he come to the States?

Mostowski Probably as a child.

Crossley What did you say, C.-C.?

Chang I said I knew that. Tarski used to tell this story in the early fifties.

Keisler What about Steve? Did that mean that Tarski was saying you had a connection with Poland?

Kleene I do not have any connection with Poland. What have I contributed to propositional calculus? Unless maybe to intuitionistic propositional calculus – which Tarski was not thinking about.

Nerode Look, just before the war in Poland, I mean before the catastrophes fell, what was going on in Polish logic? I have the impression that it was out of propositional logic by 1938 or 1939. I had the impression that Tarski had already started up trying to get model theory going. Is that right?

Mostowski Much emphasis was on these decision problems. So that was one thing. Another thing was, you know, people were working on these strange systems of Lesniewski. So this was very fashionable. And still there was work on propositional logic.

Sacks Did people know Tarski's decision procedure for the real numbers at that time?

Mostowski — Tarski gave a lecture about it, but I do not think it was generally known. He did not publish it until much later (Tarski and McKinsey [48]).

Keisler — I think there was an attempt to publish it that was interrupted by the war.

Nerode — You mean the proof sheets?

Sacks — Yeah.

Nerode — Isn't there some story about the proof sheets being sent ...

Chang — Except the originals were destroyed.

Kleene — Which proof sheets were these?

Chang — The decision method for real closed fields.

Mostowski — Was it not so that he spoke about it at the Paris Philosophical Congress? And that these proof sheets were these?

Keisler — I think that was it.

Nerode — No, the type was destroyed, so it actually never appeared, though he had already corrected the proofs.

Chang — Though he has copies – at least one that I have seen.

Nerode — I remember that too.

Chang — I guess one of the saddest things in logic is that Presburger never got his degree.

Nerode — That Presburger never got his degree?

Mostowski — Yes, yes. Tarski refused to give him a degree for his paper (Presburger [30]).

Nerode Which is now one of the most cited.

Kleene Why?

Mostowski Because he considered it too simple. He thought that it was not what he wanted.

Sacks What was this? The decidability of ...?

Nerode Addition and multiplication – separately.

Mostowski No, without multiplication.

Nerode Separately – didn't he do both?

Mostowski No, he did only addition.

Kleene How you can do it with multiplication without addition I do not know, because the usual recursion equations for multiplication have addition in them.

Nerode Sure, but I thought that Presburger had also done the theory of multiplication alone. That was my memory of it. Were you the first to do that?

Mostowski I think Skolem was the first to do that.

Nerode Ah, Skolem!

Mostowski Yes, but I think Skolem was later than Presburger. I am not sure whether it was so. But in Presburger's paper there is only addition.

Sacks So Tarski thought the proof was too simple.

Mostowski — Yes. This was a very obvious application of elimination of quantifiers. You know at that time ...

Kleene — Presburger – nine pages in 1930.

Chang — A nine-page thesis!

Mostowski — Well, he could have expanded it a bit.

Sacks (to Mostowski) — What were you saying?

Mostowski — At that time the method of eliminating quantifiers was pretty well known. Tarski took it over from Skolem, but he taught it at the university and applied it to several problems.

Sacks — I did not know that. The method of elimination of quantifiers was invented by Skolem?

Mostowski — Yes.

Keisler — What did Skolem use it for?

Mostowski — To prove a decidability result ... You know there is a paper of Skolem which has a very long name: "Untersuchungen über die Axiome des Klassenkalkuls und über Produktations- und Summationsprobleme, welche gewisse Klassen von Aussagen betreffen". That is the title.

Sacks — That is a title!

Mostowski — In effect he proves there that, if you have unary predicate calculus with quantifiers not only for individual variables but for functional variables, then this gives you a decidable theory.

Sacks What is that, second order monadic?

Mostowski Yes, second order monadic.

Sacks I did not know that was his result.

Mostowski There he uses elimination of quantifiers.

Keisler What is the year on that?

Mostowski The year is 1919.

Sacks 1919!

Nerode Did not Lewis and Langford (Lewis and Langford [32]) come out even earlier? Langford was a very early person, so he must be in there somewhere.

Keisler It must be there ________ (with reference to IM).

Kleene This is not a complete bibliography, you know. I only put a paper in there if I had some reason to cite it in the book.

Mostowski I am not certain about the year, but there is a paper of Langford involving ordering where he used more or less the same ...

Nerode It is later, it is 1927.

[General agreement.]

[Bibliographical note: Langford [26]- dense order;
[27]- discrete order.]

Mostowski At any rate since the method was known it was quite easily applied to these additions.

Chang What was Lindenbaum's thesis about?

Mostowski Lindenbaum's thesis was about metric spaces and had nothing to do with logic. He wrote this with Sierpinski. I do not know exactly what it was about, but it was something to do with the classification of metric spaces.

Crossley When was that?

Mostowski Somewhere in the middle of the twenties.

Chang So Lindenbaum was not technically a Tarski student?

Mostowski That is right.

Chang But he worked closely with Tarski.

Nerode Well, why did ... I never understood the history. Some people say Lindenbaum Algebra and others say Lindenbaum-Tarski Algebra.

Mostowski You know, the name Lindenbaum Algebras, I think I am partly responsible for that. Lindenbaum wrote a paper in which he showed that if you have a system of propositional logic, then you can find a matrix which is satisfied precisely by the theorems of this propositional logic. So, after the war, when Rasiowa and Sikorski wrote their book (Rasiowa & Sikorski [63]) and they wrote their papers and they asked how they should call the algebras built of propositions, we decided in joint conversations that because the first idea that an algebra might be an algebra constructed out of propositions was Lindenbaum's, they should be called Lindenbaum algebras. But Tarski had made an algebra built of propositions. For every theory built on the usual propositional logic he had constructed explicitly a Boolean algebra of these propositions, so he claims that it was his algebra because it was really the Boolean algebra as constructed by him. But I think the general idea as far as I understand was first conceived by Lindenbaum because Lindenbaum was the first who had this matrix. You know you should have asked Tarski. Maybe the Lindenbaum construction was suggested to him by Tarski ... that should be known. We only knew that it was Lindenbaum's theorem.

Nerode — But was there not also a controversy over Stone's theorem? I mean the representation theorem for Boolean algebras?

Keisler — You mean whether Tarski claimed it?

Nerode — Yes.

Keisler — No, Tarski was just one step behind. You mean in the construction of ultra-filters for set algebras, and he proved their existence but just missed that key step, that ultrafilters exist just for general algebras.

Mostowski (to Keisler) — How do you call such algebras – Tarski algebras or Lindenbaum algebras?

Keisler — Lindenbaum algebras, I think, yes.

Chang — There are other things here, you know, in that I think sometimes we called it the Löwenheim-Skolem-Tarski theorem, especially when it goes up, though sometimes when it goes both up and down we refer to it as the Löwenheim-Skolem theorem.

Keisler — Do we? It is supposed to be Löwenheim-Skolem-Tarski all the way through.

Nerode — Does anyone know anyone who has ever met Löwenheim; I mean, he published that one paper (Löwenheim [15]) and vanished.

Morley — Tarski.

Nerode — Tarski knew Löwenheim?

Morley — Tarski claims to be the only logician that met him.

Kleene — Skolem never met Löwenheim?

Morley — Tarski claims that he met Löwenheim once and he is the only sort of official logician that ever met him.

Kleene — Could well be.

Nerode — Did he say what Löwenheim had done for a living, so to speak?

Morley — High-school teacher or something.

Kleene — And Löwenheim of course had a very strange (from our point of view), a very strange way of putting it.

Mostowski — To me the name "Downward Upward Löwenheim-Skolem theorem" is extremely strange because Skolem-Löwenheim proved it only for the case from infinite to denumerable (Skolem [20]), and upward I think it was Tarski who proved it (Tarski [34]).

Nerode That is right.

Keisler That is why we carried the name Tarski along, but it is very awkward, I think, a three-name theorem.

Crossley Yes, there is this beautiful thing that C.-C. did in Leicester when he talked about this theorem and gave it its full title, and it was the Gödel-Henkin-Mal'cev-Feferman-Tarski-Rasiowa-Sikorski- and so on, you know, it went on ...

Morley Did you do something like that?

Crossley You know everyone who was involved with the compactness theorem plus the Löwenheim-Skolem type theorems was listed. It, in fact, was incredible.

Chang Did I do that?

Crossley You did that.

Chang That was some years ago.

Sacks Yes, you used to kid around a lot.

Nerode Steve, when did you first find out that there was a real connection between your work on function quantifiers and descriptive set theory? In other words, there is no evidence that you had looked at descriptive set theory in your 1955 paper (Kleene [55]) at all or that you saw any connection. Was that John Addison?

Kleene — John Addison was doing it already with number quantifiers, wasn't he?

Nerode — Yes.

Kleene — I think there is a footnote in my 1943 paper "Recursive Predicates and Quantifiers" (Kleene [43]) saying there is obviously some kind of parallelism here, and it was John who looked into it and checked it out. But I at that time had never read the Polish descriptive set theorists. I just knew there was such a theory. I had not looked up the papers and read them. I only did that when I was checking Addison's thesis. So Addison had no help from me in any detail on charting out what he did. It was his work.

Crossley (to Mostowski) — You were saying earlier that you were aware of this connection.

Mostowski — I was saying I was aware of some connection between some facts about functions on the integers and set theory. I considered certain things which happened here and happened there. But of course the analogy was very imperfect. A more perfect analogy was in Addison's dissertation (Addison [54]), which ...

Kleene — Addison was in Poland for a year, wasn't he?

Mostowski — Yes.

Kleene After he got his dissertation. It was very nice of you to ask him over, and I am sure he profited from it a great deal. Maybe Addison got started by this footnote in my 1943 paper, because he was looking for something to work on. And he said: "Well, here is this footnote that says 'there should be analogies', so let us look at them in detail". Also, there was something which you pointed out – that the analogy did not work. I have a paper on the symmetric form of Gödel's theorem (Kleene [50]), and I think this went counter to the analogy. Maybe that is what started Addison off, because he looked into it and he took the matching differently and made the analogy go through by a different matching.

Mostowski I remember your paper about this symmetric problem. You showed that one cannot pursue the analogy between the number-theoretic recursion theory and descriptive set theory. Only later the true analogy emerged.

Crossley I find it very difficult to get the historical perspective, the time of publishing a lot of these papers. Especially over the war period. What was the situation then? Were people in a situation where they had a lot of things they wanted published, but did not get published until after the war? You had papers which refer back six or seven years before.

Kleene The arithmetical hierarchy, as I eventually published it in "Recursive predicates and quantifiers", I found in the spring of 1940. And at that time I started writing

"Introduction to Metamathematics" and you know, there was a problem; how was I going to get a book out that people would be interested in buying? So I said: "By gum, I'll put these new results into it". So I did not start writing them up for publication as a separate paper because I was going to work them into I.M. The writing of I.M. turned out to be an enormous task, and then in May of 1942 I joined the Navy and had no more time for such things. I guess before that, before I joined the Navy, I suddenly realized that it was foolish to hold the hierarchy for a book; perhaps I should get it out as a separate article. So the discovery dates from 1940 and I think as an abstract it was published in 1940. But the paper took a couple of years to get out. As to realizability, we pretty well knew what we were getting before the war, but I waited to publish it until David Nelson had finished his part of the work and that meant that that did not get published until 1945. But I had it in manuscript before the war – modulo final revisions and references.

Keisler What about people interned during the war?

Mostowski I think many of their papers, most of them, were destroyed during the war. I wrote my paper (Mostowski [47]) on this hierarchy more or less duplicating what Steve did, after the war, but I had notes dating back to 1942-1943. Only I had a nice, very big, wonderful notebook with all these discoveries – and then in 1944 there was an uprising in Warsaw and I remember the soldiers came to our house and ordered us to leave. So I was with my mother in this house and I hesitated whether I had to take the notebook with me or some bread. I decided to take some bread, so all my notes were burnt. So then I reconstructed the papers some time in 1945.

Chang I think you told me you worked on tiles – on roof tiles, during the war.

Mostowski Yes, I worked in this factory where they made these tiles. I did not have very much to do with that, so I devoted most of my time to logic.

Kleene I simply closed the books on logic in May 1942 (except for publishing in two papers some work already done) and did not reopen them until – well, it might have been November - October or November 1945. Because I did have a lot to do and I felt it my duty to do the most I could in that war, I thought.

Mostowski I will tell you still one story. After the soldiers came and I took my bread ... we were led to what the Germans call a "Durchgangslager", that is the term. And people were divided in two parts. The older ones in one direction, the younger in the other direction and the younger ones had to be sent to a concentration camp, but I somehow escaped – but that is beside the point. At any rate I met there a professor and if you can imagine a group of such people carrying bags with a large group of soldiers around us, and I see the professor And he greets me very kindly and I ask him:

"What are you carrying in this bag?" "Oh", he said, "I am carrying the manuscript of my book which I wrote during the recent months". "What is the book about?" I asked. "The book is about happiness". And the book was published because he somehow carried it.

Nerode That is marvellous. Tell us about that escape. The one you just admitted. You said that you had managed to get away from this.

Mostowski There were nurses, Polish nurses, there; I knew some of them – and they led me through the German line to a hospital and then a doctor gave me his identity card which permitted me to go out posing as him.

So that was my first escape. Then there was another one – and a third one because the Germans did not like people with documents from Warsaw, and they used to arrest us.

Keisler Where were you between escapes?

Mostowski First I stayed with my relatives who were living near Warsaw and then I went to look for a colleague of mine who owned a house – also not very far from Warsaw. And then I worked on a farm.

Sacks And then the Russians came ...

Mostowski And after several months the Russians came. Before they came I – well, you know people helped each other – so there was a branch of the agricultural university which was already occupied by the Russians and the rest of it was transferred to the west of Warsaw. So my wife and I went there and asked to be admitted as former workers of this institute - which was absolutely false and everybody knew that this was false, that we were just posing as former employees of this institute. But they accepted us because they wanted to help us. So we got documents that we are former employees and then the Germans stopped harassing us, and we lived more or less quietly until the Russians came.

Keisler Why were you harassed in the first place? Was it because you were an intellectual?

Mostowski Because I was from Warsaw. You know the population of Warsaw was expected to be either sent to Germany to work or to special camps where older people were waiting until the war ended. So I had a document showing that I am a permanent inhabitant of Warsaw and these people were prosecuted, they were not allowed to leave without special permission. This was during the last couple of months after the uprising of Warsaw. (Pause) But that does not belong to the history of logic.

Crossley No, but the history of logic would be different if you had been eliminated at that time.

Nerode That is true.

Kleene Much different.

Mostowski — You know, I always thought that my history would be different if I had decided to take my notebook with me instead of bread.

Morley — Yes, you really had great theorems there which have never appeared ...?

Mostowski — No, but you know, very useful theorems, for instance, I could reconstruct all the proofs which Gödel had about consequences of $V = L$, for descriptive set theory. These were later published by Addison (Addison [59]) and Novikoff (Novikoff [51]). But I had it already then and I had also the decidability of the elementary theory of well-ordering. There is an abstract of Tarski and myself about it (Mostowski and Tarski [49]), but the proof was never published because the procedure of eliminating quantifiers was hard and required too many details to be worked out, and I had this all written up there, I regret that.

Nerode — How did you go about re-establishing Polish logic immediately after the war?

Mostowski — The University started working, well, as a matter of fact we had teaching during the occupation although the University was closed. There was an illegal organisation, I mean illegal in the sense that the Germans did not allow us to continue that, but teaching was going on, so after the war when the University was opened and teaching started we had already students who were trained a bit during the occupation.

Nerode When I was educated, you know the 1950's, we were told that the Russians regarded most of mathematical logic that time as basically anti-Marxist and after the Russians took over really control of Poland it would have seemed to me very difficult to pursue logic, and evidently that is not what happened.

Mostowski I do not think this may be true, though I believe there must have been several philosophers who looked upon mathematical logic with suspicion. I remember I spoke with Kolmogorov, who was the boss of the whole of mathematics, and he invited me to come to Moscow and to lecture about logic (which I did not do). I think there was a sharp division between mathematicians who regarded logic as something quite admissible, quite, how shall I say, respectable, and philosophers who looked upon it as with suspicion.

Crossley That is interesting, because at the time I was a graduate student there was a great deal of suspicion about mathematical logic from other mathematicians.

Sacks That was in England.

Morley That is correct.

Kleene This is a little bit different from political disfavour. Just, you know, it was a matter of respectability.

Nerode No, but they were equally suspicious of algebraic topology.

Kleene Yes.

Sacks I did not know that.

Kleene There was a time when, in this country, there were many places when a mathematical logician was not really considered a respectable mathematician.

Crossley You mean this country?

Kleene The United States.

Crossley The United States.

[Laughter]

Kleene Did I ...?

Crossley Yes, you did!

[Laughter]

Mostowski I believe that all mathematicians ...

Sacks I am still waiting for one to get into Columbia, though.

Mostowski ... look upon logic with suspicion. But I think if you have a famous man like Kolmogorov who is doing logic and who is respecting logic, then you can expect a number of mathematicians will say: "That is a strange field but we must accept it because ...". And I think this was more or less the situation in Moscow.

Nerode Do you know how Mal'cev arose? In other words, here he is in the 1930's as the only one trying to do logic in a serious way in the Soviet Union, or at least so it appears to us on the outside. But who was his adviser, and how did he get into logic?

Mostowski He taught in Ivanovo, which was a very small school. (Pause)

Keisler Has there ever been any axiomatic set theory done in Russia?

Mostowski No, I don't think so. I don't think so. You know, there were very clever people in Russia. Lusin, first of all, was there who was interested in problems of descriptive set theory, which bordered the foundations of mathematics. Of course he was more close to the French intuitionists and semi-intuitionists, so he probably taught them set theory, not axiomatic set theory, but set theory and foundational problems.

Keisler Even to this day I do not know of any axiomatic set theory that has been done there. Is there any reason or is it just coincidence?

Mostowski This must be the influence of Lusin who, you know, was more or less a contemporary of Lebesgue and Baire and he took it all very intuitively.

Sacks Was Mal'cev located in Moscow?

Mostowski I think he must have studied in Moscow. But when I first heard of him he was a professor at an Institute at Ivanovo which is some five hundred miles or one thousand miles to the east of Moscow, so he was very isolated.

[Pause]

Nerode Steve, did you know Turing when he was in Princeton?

Kleene I never met Turing.

Nerode You never met Turing?

Kleene No. Once I talked to him on the telephone. You see, I left Princeton in June 1935, and I do not know whether he came in the fall of '36, or the fall of '37; anyway, it was after his paper was out and the Princeton people invited him, but I was not in Princeton then. I was back in Princeton in 1939-40, but by then he was gone. I used to go to the summer meeting (early September meeting) and also to the Christmas meeting of the Mathematical Society, and I guess Turing just did not go to these things. Then one day in July 1950 I turned up in Leicester, England, where my wife had relatives – and I could have written Turing ahead of time but I did not – I turned up in Leicester and I telephoned Turing in Manchester, which was, you know, oh, an hour or two on the train or something like that, and offered to come over and see him. And he said: "Well, don't come because I am tied up with something- it is important to me to run something on the machines", or something like that. I felt that it was partly shyness, because, you know, I thought he could have found maybe some time to talk to me. Anyway, that is the closest I ever got to meeting Turing. I did not get to Europe to meet logicians till 1948. I was once there when I was an undergraduate student in 1929, but I did not get to Europe again till 1948, and Gentzen was killed in the war, so I missed meeting Gentzen. I never met Hilbert. Most of the rest I met. Well, I am sure you will find some Poles I did not meet!

Mostowski Yes.

[End of tape]

[We had hoped Robin Gandy would be present to talk about Alan Turing but in the course of the meeting it turned out that Peter Hilton had known Turing. So on 29 January 1974 there was another session at which John Crossley, Peter Hilton, David Lucy, Andrzej Mostowski and Liz Wachs-Sonenberg were present.]

Mostowski What I remember I was telling you was that I attended a lecture of Gödel and I think this was the first publication, if one can call a lecture a publication, of the theory of his result on the consistency of the axiom of choice. So he had a one-semester course in Vienna on axiomatic set theory in which he gave axioms for what is now called Gödel-Bernays set theory, and then developed a model. He constructed a model in which the Axiom of Choice was valid. At that time, I am sure that he did not have the consistency proof for the continuum hypothesis, because he restricted his lecture exclusively to the axiom of choice and the construction went more or less like this: he had these levels of the constructive hierarchy, defined, more or less, as he defined them later in his paper on consistency of the continuum hypothesis, but he did not formulate the axiom of constructibility. Only he proved that all these levels have well-orderings, so that a well-ordering on a given level can be lifted to the well-ordering of the next level and also on limit ordinals you can get this well-ordering. And so in this way he obtained a model in which each set was well-ordered within the model. That was his construction. He never mentioned, at that time, that he had a proof for the continuum hypothesis. So I must say after I read his publication – it was a few years after that, that he published this short paper in the American Academy of Science – I was very upset that he carried this work so much further, because at that time he had only this very weak result.

Crossley I think one area which we did not cover at all, which I think is perhaps most relevant to this context, was how you got into recursion functions. When did you start working on recursion function theory?

Mostowski I got to recursive functions – I think this was due to the paper of Kleene. I studied his paper (Kleene [36]) in the Mathematische Annalen and was struck by the analogies between his construction- his theory- and the theory of projective sets. There are certain analogies. So I was very much interested that there are these analogies. I think this was quite independent of my main work, you know, because I was mostly interested in the axiom of choice at that time. I wrote a Ph.D. thesis on the axiom of choice, on models which you called Fraenkel-Mostowski and also I wrote a paper (Mostowski [48]) on the various definitions of finiteness, which was again connected with Gödel's incompleteness theorem. The question of Tarski was whether there exists a weakest definition of finiteness. He believed that this Russell definition is the weakest possible definition of finiteness. And I was able to show, using Gödel's results, that there is no such weakest definition. And since I was at that time living in Zürich, I was quite alone, I knew nobody and I was interested in various things, and I read the paper of Kleene which I liked very much. Then during the war I worked out some analogies between the theory of arithmetical hierarchies and projective hierarchies. Of course they were very superficial, the analogies, but still there were some. For instance, if you have a recursively enumerable set whose complement is recursively enumerable it is recursive and in projective hierarchies you have the same for Σ_1 sets and analytic sets. So I remember that I worked all of this out during the occupation and published it afterwards. So this was completely independent of my main line of development.

Crossley So you did not have much significant contact with computability as such?

Mostowski No. So I must say that I never understood this connection between computability, and Church's Thesis was completely unclear to me and it took me many years to understand what it is really all about. So I took the theory as it was developed by Kleene. I considered it to be a very nice mathematical theory, but was not interested it its connections with computability. These analogies were very charming to me. I think they were later developed in a much better way. I think this work was more attractive to me than most.

Crossley Whom did you do your thesis under?

Mostowski You know, you must differentiate between practice and theory. In fact it was Tarski who was my supervisor, but since he was not a professor, Kuratowski acted in his place. So in the documents it is written that he was my thesis adviser, I do not know what the English word is for that - but in fact it was Tarski who was the actual thesis adviser. I spoke very much with Lindenbaum at that time.

Crossley With whom?

Mostowski Lindenbaum, who is not very well known, I think, in the world, but he was one of the most intelligent men in logic at that time in Warsaw. And he told me that I should read the paper of Fraenkel [22] and make this method of Fraenkel more precise. We discussed this problem and we published a joint paper (Lindenbaum and Mostowski [38]) about it. I think he had a great influence on the development of my ideas on the axiom of choice.

Crossley So how did you come to be in Zürich?

Mostowski I believe every student who is interested in mathematics should travel a bit after his work is done, so I had some means to go abroad and I went first to Vienna and then to Zürich. I thought that I would study applied mathematics in Zürich, which was thought at that time to be a centre of actuarial mathematics. And since jobs were very scarce at that time in Poland, I thought it would be necessary for me to devote myself to some practical work. I tried to study this statistical business, but I found the lectures so boring that I gave it up.

Hilton Was Saxer there then?

Mostowski I do not remember the name. I remember there was a special lecture for actuarians. Actuarians?

Hilton Actuaries.

Mostowski It was terribly, terribly boring – a very elementary kind of mathematics was used in it. But I attended the classes of Polya, and his seminar and also a seminar of Bernays. And these were extremely interesting. So nothing came out of my efforts to get practical – to get training in applied mathematics and I worked all the time on recursive functions and on the axiom of choice.

Crossley Did you ever meet Hilbert?

Mostowski No. Hilbert was in Germany at the time. Really the first time I was in Germany was in the fifties, after the war. But I saw Gentzen – I think I even spoke a few words to him. It was in 1937. There was a Congress of philosophers in Paris. It was an international exhibition – one of the first of those fairs or exhibitions in Paris. There was also a philosophical congress and Gentzen spoke there about Hilbert's programme and announced his results about the epsilon number and what was the possibility of proving consistency and it was very impressive.

Crossley I get the impression that you met perhaps fewer of the logicians than Steve Kleene did at that time. Is that true? Or is it just that there were a lot of Polish logicians about?

Mostowski Oh, certainly I met much fewer of them, you know. We had, of course, quite a strong school in Poland, and I knew quite a few of them, but from abroad – well, I don't remember having met very many people. So I knew Gödel and I met Södermann, the Finnish logician who died very early during the war, or after the war, and a couple of other people – Hermes for instance – but I did not have much contact with them. The language barrier was very strong.

Crossley We had hoped to get Robin Gandy to talk about Turing, but I just discovered the other night that Peter had known Turing, so we thought we should get something about Turing from Peter. That should virtually complete this sort of survey.

Hilton Well, I cannot talk, of course, with any authority about his work in logic.

Crossley It is just to get some impression of what things were like at that time. When did you first meet Turing?

Hilton — I first met Turing on January 12, 1942.

Mostowski — What time?

[All: laughter]

Hilton — I could almost tell you that, but that was the date on which I began my war service and I had no idea what I had been selected to do, because I had been interviewed in Oxford where they were looking for a mathematician with a knowledge of German. And I did not have to be any good at either of them because there was nobody else. And so I was chosen and I had to present myself on January 12 at Bletchley, Buckinghamshire, and then I went along and this man came over to speak to me and he said: "My name is Alan Turing. Are you interested in chess?" And so I thought: "Now I am going to find out what it is all about!" So I said: "Well, I am, as a matter of fact." He said: "Oh, that is very good, because I have a chess problem here I can't solve".

And in fact he just wanted to think about the chess problem. And this was typical of Turing, that he had these tremendous passions. It might be for the time being chess; or it might be Go, or it might be tennis. But whatever it might be, for the time being that was what he was on about. And it was in fact 24 hours after I first met him, that I discovered what in fact I was there to do.

Everybody called him "Prof", which was rather interesting, because in England, the title "Professor" of course, as here in Australia, is not simply a mark that you are in the academic profession, but it is a mark of your having a certain rank in that profession, which of course Turing in 1942 had not had. But nevertheless there was this feeling about this man that he was the "Prof". I remember after the war this created a little bit of embarrassment because those who had known him during the war always continued to call him Prof, and he was still not a professor in the strict technical sense. And some of the actual professors did not like the fact that he was always referred to as Prof, and if the word Prof was used without a name it always meant Turing.

He was a very easily approachable man – though you always felt there was lots more you did not know anything about. There was always a sense of this immense power and of his ability to tackle every problem, and always from first principles. I mean, he not only, in our work during the war, did a lot of theoretical work, but he actually designed machines to help in the solution of problems – and with all the electric circuitry that would be involved, as well. In all these ways he always tackled the whole problem and never ran away from a calculation. If it was a question of wanting to know how something would in fact behave in practice, he would then do all the numerical calculations as well. And, of course, as you know, he designed the computer at the National Physical Laboratory and also the Manchester computer which were the first computers, except for the one at Cambridge, to be working in England after the war. So he was certainly quite an inspiration to us and a very lovable man.

I say there were strange features about him. For instance he had this extraordinary idea that money would lose its value in England, win or lose the war. And consequently, that he should convert what little money he had into silver bars and bury them and then dig them up again afterwards, when they would of course regain their real value. And he proceeded to do this, and if my recollection is correct he told us that he had buried them on Salisbury Plain, and he had very carefully noted the coordinates inside this forest of the point where he had buried these silver bars. But he had the misfortune that, when he returned, I think in 1944, with a mutual friend, Donald Michie, the whole terrain had been changed. The forest was down. There were houses and so forth, and he never did in fact find these silver bars. But he kept returning to what he thought was the spot and, again characteristic of the man, he built his own metal detector. And this metal detector was remarkable, as Donald Michie told me, because it simply changed the pitch of the note it emitted in the presence of metal. So you had to tolerate a continuous noise which would just change. But it worked – it worked. As with all things with Turing, it really did work.

Max Newman has written a very appreciative obituary

(Newman [55]) of Turing in the Obituaries of Fellows of the Royal Society in which he refers, for example, to Turing's bicycle, which was a very famous machine at Bletchley. Only he could ride it without the chain falling off, because he knew that if he rotated the wheels at a certain speed and back-pedalled very sharply at a certain time interval, he could avoid the chain falling off when a certain missing ratchet was just about to come into contact with the chain. Nobody else could ride it. And he said: "This is much more efficient than having the thing repaired, because I don't have to buy a lock. It is well known that nobody else can ride it".

There is another very nice story of Turing that he was a civilian, working in Intelligence, and he believed – again typical of Turing thinking in first principles – that the Germans might very well invade England and that then he should be able to fire a rifle efficiently, and so he enrolled in what was called the Home Guard. The Home Guard was a civilian force, but which submitted to military training and in particular its members learnt how to fire a rifle. (They might have learnt some other things, they might have been in the radio section or something of this sort. Turing, in fact, enrolled in the Infantry part.) In order to enrol you had to complete a form, and one of the questions on this form was: "Do you understand that by enrolling in the Home Guard you place yourself liable to military law?" Well, Turing, absolutely characteristically, said:

"There can be no conceivable advantage in answering this question: 'Yes' ", and therefore he answered it 'No'. And of course he was duly enrolled, because people only look to see that these things are signed at the bottom. And so he was enrolled, and he went through the training, and became a first-class shot. Having become a first-class shot he had no further use for the Home Guard. So he ceased to attend parades. And then in particular we were approaching a time when the danger of a German invasion was receding and so Turing wanted to get onto other and better things. But of course the reports that he was missing on parade were constantly being relayed back to Headquarters and the officer commanding the Home Guard eventually summoned Turing to explain his repeated absence. It was a Colonel Fillingham, I remember him very well, because he became absolutely apoplectic in situations of this kind. This was perhaps the worst that he had had to deal with, because Turing went along and when asked why he had not been attending parades he explained it was because he was now an excellent shot and that was why he had joined. And Fillingham said: "But it is not up to you whether you attend parades or not. When you are called on parade, it is your duty as a soldier to attend". And Turing said: "But I am not a soldier". Fillingham: "What do you mean, you are not a soldier! You are under military law!" And Turing: "You know, I rather thought this sort of situation could arise", and to Fillingham he said: "I don't know I am under military law". And anyway, to cut a long story short, Turing said: "If you look at my form you will see that I protected myself against this situation". And so, of course, they got the form; they could not touch him; he had been improperly enrolled. So all they could do was to declare that he was not a member of the Home Guard. Of course that suited him perfectly. It was quite characteristic of him. And it was not being clever. It was just taking this form, taking it at its face value and deciding what was the optimal strategy if you had to complete a form of this kind. So much like the man all the way through.

We were all very much inspired by him, his interest in the work but the simultaneous interest in almost everything else. As I say, it might be chess, it might be Go, it might be tennis and other things. And he was a delightful person to work with. He had great patience with those who were not as gifted as himself. I remember he always gave me enormous encouragement when I did anything that was at all noteworthy. And we were very very fond of him. And then after the war he went to the National Physical Laboratory and our group dispersed. But in fact I again resumed my friendship with him because he went from the National Physical Laboratory to Manchester. I do not remember exactly when, but I should suspect it was 1946 or 1947 – but I am not sure. And I went to Manchester in 1948. Newman, who had been again one of the people at

Bletchley, had invited Turing to come to Manchester because of Newman's interest in developing the computing machine and Turing came as a Reader, in the technical English sense of Reader. I went as Assistant Lecturer, my first academic job. Once again, I found Turing as delightful as ever, and he explained to me about machines and what he saw as their possibility. At that time, discussion about computers was very largely in the hands of educators and bishops and such people who were objecting very strongly to the use of anthropomorphic terms like "think" when applied to computers – I mean, something which now seems terribly old hat but at that time was very controversial. I remember very well Turing being engaged in debate on a programme on the BBC, "Can Machines Think?" and I believe it was in fact a bishop (certainly it was a cleric) who was taking the opposing view and objecting that machines could only do what they were instructed to do. And Turing said: "Really, this is very unfair. You are apparently complaining because we have built deterministic machines. Now, it would be perfectly simple for us to put a randomizing element into the machine if we wanted to and make the behaviour quite unpredictable, but simply the purposes for which we have designed the machine would make that a silly thing to do. "But", he said, "this presents no problem at all. In fact", Turing added – and I could just see the sort of malicious smile he would have had on his face when he said it – "I can envisage, in the future, two ladies wheeling their computers in the park and stopping to talk to each other and one saying to the other: "Do you know? My little computer said such a funny thing to me this morning"." This did not go down well with the bishop but Turing very much liked to shock in that respect. He was a boy in that way. He enjoyed that.

Another recollection I have from that period, which is a little closer to his work in logic. I remember it surprised me intensely at the time, because, of course, his stature as a mathematician was immense, and I was a raw beginner. Sonehow or other, in the Mathematics Department at Manchester the question of the solvability of the word problem in groups came up and Turing claimed he had never heard of this problem and found it a very interesting problem, and so, though at that time his principal work was in machines, he went away and about ten days later announced that he had proved that the word problem was unsolvable. And so a seminar was arranged at which Turing would give his proof. And a few days before the seminar he said : "No, there was something a little wrong in the argument, but the argument would work for cancellation semigroups". And so he in fact gave his proof for cancellation semi-groups.

What may be of some small interest as well to an Australian audience was that, if my recollection is correct, Bernhard Neumann also said that he had not heard of the problem and believed that he could show that, for groups, the word problem was solvable. But he also withdrew that before he was put on the rack. I remember being very surprised because certainly I, as a rank novice, had heard about this problem, probably from Henry Whitehead.

Well, I don't know, John. Is that really enough? I mean, there are other little anecdotes connected with Turing. There is also the very tragic circumstances connected with the business of his death. I remember the terrible shock. It was in 1954, I think. I had recently gone from Manchester to an appointment at Cambridge and within a short period of my being there, if my recollection is correct, I heard the news of Turing's death. It was a sense of terrible loss. All those who knew him had a great affection for him.

Mostowski But was his work in this higher Intelligence unit of the army successful?

Hilton Yes, yes.

Mostowski So what did you do? If that is not a classified ...

Hilton It is all right to say now... We were engaged in cryptography. I believe it is all right to say that, so I shall risk it, because books have now been published which talk about this as having gone on at Bletchley. And Turing was an absolutely key figure, not only in developing the broad methodology of the attack, but also in the detailed work.

Hilton (continued)

But there again, you see, he began to be beset by the bureaucrats who wanted him to come in at a certain time and work till five o'clock and leave. His procedure — and that of many others of us, let me say, not only he, who were really fascinated by the work — would be maybe to come in at midday and work until midnight the next day. And then, the problem being essentially solved, go off and rest up and not come back for 24 hours perhaps. But they were getting much much more work out of Alan Turing that way. But, as I say, the bureaucrats came along and wanted forms to be filled in and wanted us to clock in, and so on.

Crossley

Where was Turing before he went to Bletchley?

Hilton

He was at Cambridge, at King's College, Cambridge. He was a Fellow of the College. I do not know whether it was a tutorial fellowship or a research fellowship. You can find that in Newman's obituary.

This is funny. I find it hard to realize how young he was. He was born, I think, in 1911. So he was only 31, maybe 30, when I first met him.

Lucy

1912.

Hilton

He was born in 1912, you say? So in fact he was 29 when I met him. I find that impossible to believe, that he was only 29. because of the immense stature that he had at that time already ...

Lucy

It is in the Gandy-Yates volume, the date. It is

dedicated to him.

Hilton It is also quite interesting to recall, in view of what we have heard at this conference on one of the most exciting present-day areas of application of mathematics, that his last paper of all was on morphogenesis. That was the area that was exciting him in the last year of his life.

Crossley Thank you very much.

* * *

BIBLIOGRAPHY

ACZEL, P. and P.G. HINMAN

[74] Recursion in the Superjump, in Generalized Recursion Theory, North-Holland, 1974, 3-42.

ADDISON, J.W.

[54] On some points in the theory of recursive functions, Dissertation, University of Wisconsin, Unpublished.

[59] Some consequences of the axiom of constructibility, Fundamenta Math. 46, 337-357.

CHANG, C-C. and H J KEISLER

[73] Model Theory, North-Holland, 1973.

CHURCH, A.

[36] An unsolvable problem of elementary number theory, American Journal of Math. 58, 345-363, 1936.

DAVIS, M

[65] The Undecidable, Raven Press, 1965.

FRAENKEL, A.A.

[22] Der Begriff "definit" und die Unabhängigkeit des Auswahlaxioms, Sitz. Preuss. Akad. Wiss., Phys.-math. Kl. 253-257. English translation in van Heijenoort, 'From Frege to Gödel, A source book in mathematical logic, 1879-1931', Harvard Press, 1967.

GÖDEL, K

[30] Die Vollständigkeit der Axiome des logischen Funktionenkalküls, Monatshefte für Mathematik und Physik 37, 1930, 349-350. English translation in van Heijenoort [67].

[31] Über formal unentschiedbare Sätze der Principia Mathematica und verwandter Systeme I, Monatsh. Math. u. Phys. 38, 1931, 173-198. English translation in van Heijenoort [67], 596-617.

GÖDEL, K.

[31-32] Remarks contributed to a "Diskussion zur Grundlegung der Mathematik", Erkenntnis, vol. 2, 147-148.

[34] On undecidable propositions of formal mathematical systems, Notes by S.C. Kleene and Barkley Rosser on lectures at the Institute for Advanced Study, 1934. Mimeographed, Princeton, N.J. Reprinted in "The Undecidable", 39-74.

[36] Über die Länge von Beweisen, Ergebnisse eines mathematischen Kolloquiums 7, (for 1934-5 publ. 1936, with note added in press), 23-24.

[38] The consistency of the axiom of choice and of the generalized continuum hypothesis, Proc. of the Nat. Acad. of Sciences 24, 556-557, 25, 220 - 224, 1938.

[46] Remarks Before the Princeton Bicentennial Conference on Problems in Mathematics, in "The Undecidable", 84-88, 1946.

HILBERT, D. and W. ACKERMANN

[28] Grundzüge der theoretischen Logik, Berlin, Springer, 1928.

HILBERT, D. and P. BERNAYS

[34] Grundlagen der Mathematik, Vol.1, Berlin, Springer, 1934.

[39] Grundlagen der Mathematik, Vol.2, Berlin, Springer, 1939.

KLEENE, S.C.

[36] General recursive functions of natural numbers, Mathematische Annalen 112, 1936, 727-742.
For an erratum and a simplification, cf. Jour. Symb. Logic 3, p.152, 2, p.38 and 4, top p.iv at end.

[43] Recursive Predicates and Quantifiers, Trans. Am. Math. Soc. 53, 1943, 41-73.

[50] A symmetric form of Gödel's theorem, Koninklijke Nederlandsche Akademie van Wetenschappen, Proc.of the section of sciences, vol.53, 1950, 800-802,
also in Indagationes mathematicae, vol.12, 244-246.

[52] Introduction to Metamathematics, Wolters-Noordhoff and North-Holland, 1952.

[55] Arithmetical Predicates and Function Quantifiers, Trans. Am. Math. Soc. 79, 1955, 312-340.

KLEENE, S.C. and E.L. POST

[54] The upper semilattice of degrees of unsolvability, Annals of Mathematics 59, 1954, 379-407.

KRIPKE, S.

[64] Transfinite recursions on admissible ordinals, I, II, Abstract, Jour. Symb. Logic 29, 1964, 161-162.

LANGFORD, C.H.

[26] Some theorems on deducibility, Annals of Mathematics 28, 1926, 16-40.

[27] Some theorems on deducibility, 2nd paper, ibid. 28, 1927, 459-471.

LEWIS, C.I. and C.H. LANGFORD

[32] Symbolic Logic, The Century Co., New York and London, 1932.

LINDENBAUM, A. and A. MOSTOWSKI

[38] Über die Unabhängigkeit des Auswahlaxioms und einiger seiner Folgerungen, Comptes Rendus des Séances de la Société des Sciences et des Lettres de Varsovie, Cl. III, 31, 1938,27-32.

LÖWENHEIM, L.

[15] Über Möglichkeiten im Relativkalkül, Math. Ann. 76, 1915, 447-470, English translation in van Heijenoort [67].

MACHOVER, M.

[61] The theory of transfinite recursion, Bull. Am. Math. Soc. 67, 1961, 575-578.

MOSTOWSKI, A.

[38] Über den Begriff einer endlichen Menge, Comptes Rendus des Séances de la Société des Sciences et des Lettres de Varsovie, Cl.III, 31, 1938, 13-20.

[47] On definable sets of positive integers, Fundamenta Math. 34, 1947, 81-112.

MOSTOWSKI, A. and A. TARSKI

[49] Arithmetical classes and types of well ordered systems, Abstract, Bull. Am. Math. Soc. 55, 1949, p.65.

NEWMAN, M.

[55] 'Alan Mathison Turing, 1912-1954', Biographical Memoirs of Fellows of the Royal Society, Vol.1, 1955, 253-263.

NOVIKOFF, P.S.

[51] O néprotivoréčivosti nékotoryh položénij déskriptivnoj teorii množéstv, (On the consistency of some theorems of the descriptive theory of sets), Trudy Matématičeskogo Instituta iméni V.A. Stéklova, Vol.28, 1951, 279-316.

PLATEK, R.

[71] A countable hierarchy for the superjump, Logic Colloquium '69, ed. Gandy, Yates (North-Holland), 1971, 257-272.

POST, E.L.

[36] Finite combinatory processes - formulation 1, Journal of Symb. Logic 1, 1936, 103-105.

PRESBURGER, M.

[30] Über die Vollständigkeit eines gewissen Systems der Arithmetik ganzer Zahlen, in welchem die Addition als einzige Operation hervortritt,
Sprawodzanie z I Kongresu Matematyków Krajow Słowianskich (Comptes-rendus du I Congrès des Mathématiciens des Pays Slaves), Warsaw 1929, Warsaw 1930, 92-101, 395.

RASIOWA, H. and R. SIKORSKI

[63] The mathematics of metamathematics, Warsaw, 1963.

SKOLEM, T.

[19] Untersuchungen über die Axiome des Klassenkalküls und über Produktations- und Summationsprobleme, welche gewisse Klassen von Aussagen betreffen,
Skrifter utgit av Videnskapsselskapet i Kristiana, I Matematisk - naturvidenskabelig klasse 1919, no.3, 37 pp.

[20] Logisch-kombinatorische Untersuchungen über die Erfüllbarkeit oder Beweisbarkeit mathematischer Sätze nebst einem Theoreme über dichte Mengen, ibid. no.4, Engl. transl. of §1 in van Heijenoort.

SKOLEM, T.

[34] Über die Nichtcharakterisierbarkeit der Zahlenreihe mittels endlich oder abzahlbar unendlichvieler Aussagen mit ausschliesslich Zahlenvariablen, Fundamenta Math. 23, 1934, 150-161.

TAKEUTI, G.

[55] On the theory of ordinal numbers, Journ. Math. Soc. Japan 9, 1955, 93-113.

[60] On the recursive functions of ordinal numbers, Journ. Math. Soc. Japan 12, 1960, 119-128.

TARSKI, A.

[34] See Skolem's 1934 paper. Editor's note 3, p.161.

TARSKI, A. and J.C.C. McKINSEY

[48] A Decision Method for Elementary Algebra and Geometry, Berkeley, Los Angeles, 1948.

TURING, A.M.

[36-7] On computable numbers, with an application to the Entscheidungsproblem, Proc. Lond. Math. Soc. ser. 2, vol. 42, 230-265, A correction, ibid. vol. 43, 1937, 544-546.

VAN HEIJENOORT, J.

[67] 'From Frege to Godel, A source book in mathematical logic, 1879-1931', Harvard Press, 1967.

WHITEHEAD, A.N. and B. RUSSELL

[10,12, 13] Principia Mathematica, Vol.1, 1910, Vol.2, 1912, Vol.3, 1913. Cambridge, England (University Press).

* * * * *

Department of Mathematics, Monash University, Clayton, Victoria, 3168, Australia.

FRAMES AND MODELS IN MODAL LOGIC[1]

M. J. Cresswell

A better title for this paper would be: "An introduction to relational semantics for normal modal propositional logic". Some of the material goes back as far as the unpublished work of Lemmon and Scott [66]; much of it will be found in the work of Segerberg [68,71]. Nothing in the paper is original but it seems to me that a short introduction which (a) begins at the beginning and (b) concentrates only on normal modal propositional logics, might be useful. Readers who would like to know more of the genesis of the subject and the intuitive interpretation of the range of systems should consult Hughes and Cresswell [68]. Readers who would like a much more comprehensive presentation of the topics covered in the present paper may be referred to Segerberg [71].

1. *Basic Syntax and Semantics*

A language $\mathcal{L}$ for propositional modal logic consists of the following:

---

[1] The present paper covers the material in the first four of six lectures I gave at Monash in January 1974. The only other short paper I am aware of which can form an introduction to the techniques expounded in the present paper is Segerberg's [68]. Readers would be well advised to proceed to Segerberg [68] from the present paper and then tackle the more extended Segerberg [71]. An early paper presenting the completeness results of section 2 by a similar method is Makinson's [69].

l.1 A denumerable set P of *proposition letters* (sometimes called *propositional variables*). We refer to these as $p, q, r, p_1, q_1, r_1 \ldots$ etc.

l.2 The five symbols, $)$, $($, $\sim$, $\vee$, L . These are called *logical constants* and must be distinct from the proposition letters.

The set S of sentences (or wff) of $\mathcal{L}$ is the smallest set satisfying

l.3 If p is a propositional letter then $p \in S$

l.4 If α and β are in S then so are $\sim\alpha$, $(\alpha \vee \beta)$ and $L\alpha$.

We make use of the following abbreviations

l.5 $(\alpha \supset \beta)$ for $(\sim\alpha \vee \beta)$

l.6 $(\alpha . \beta)$ for $\sim(\sim\alpha \vee \sim\beta)$

l.7 $(\alpha \equiv \beta)$ for $((\alpha \supset \beta) . (\beta \supset \alpha))$

l.8 $M\alpha$ for $\sim L \sim\alpha$

You will no doubt recognize the first three as one notation (that of *Principia Mathematica*) for ordinary propositional logic. You will also know that there are many others. In place of L, $\Box$ is frequently used. In place of M, $\Diamond$ is often used. The original intuitive meaning of L was 'it is necessary that' and of M 'it is possible that'.

All the logics which will be discussed in this paper will be stated in a language of the kind defined. Trivial variants are obtained by basing the language on other primitive symbols, but in the context of the matters we shall be discussing they are not essentially different.

By a *normal modal logic* (in $\mathcal{L}$) is meant a set Λ of wff of $\mathcal{L}$ such that

1.9 Λ contains all PC-tautologies

1.10 If $\alpha \in \Lambda$ and $p_1, \ldots, p_n$ are propositional variables and β is α with $\gamma_1, \ldots, \gamma_n$ uniformly replacing $p_1, \ldots, p_n$ then $\beta \in \Lambda$

1.11 If $\alpha \in \Lambda$ and $\alpha \supset \beta \in \Lambda$ then $\beta \in \Lambda$

1.12 If $\alpha \in \Lambda$ then $L\alpha \in \Lambda$

1.13 $L(p \supset q) \supset (Lp \supset Lq) \in \Lambda$

1.9 can be summarized by saying that Λ contains PC and 1.10-1.12 by saying that it is closed with respect to substitution, Modus Ponens and Necessitation[2].

It should be clear that the intersection of any class of normal modal logics is itself a normal modal logic. We follow Lemmon and Scott [66] in denoting the intersection of all normal modal logics by 'K' (for Kripke).

For the purposes of this paper we shall look at a small number of the simplest normal modal logics. In the following list the logic in question is the intersection of all the normal modal logics containing the logics and formulae listed beside it.

T: $Lp \supset p$

S4: T, $Lp \supset LLp$

B: T, $\sim p \supset L\sim Lp$

---

[2] Non-normal logics do not always contain the rule of necessitation. For studies of these *vide* Kripke [65], Hughes and Cresswell ([68], chapter 15) and Segerberg ([71], chapter IV). A *logic* as a set of formulae is to be distinguished from an *axiomatic system*, which consists of an effectively specifiable set of axioms and transformation rules. A logic Λ is *axiomatizable* if and only if there is an axiom system $\mathfrak{A}$ such that Λ is the set of theorems of $\mathfrak{A}$. Λ is *finitely axiomatizable* if and only if there is such an $\mathfrak{A}$ with only finitely many axioms. Obviously the same logic can be axiomatized in different ways. Not all logics are axiomatizable.

S5: S4,B (or else; T, $\sim Lp \supset L\sim Lp$)

These logics are related as follows, where the arrow indicates proper containment

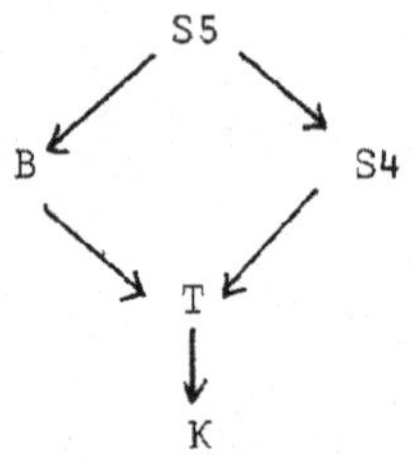

There are of course many more normal modal logics (*vide* e.g. Appendix 3 of Hughes and Cresswell [68]). All the logics we consider in this paper are normal modal logics in $\mathcal{L}$ unless explicitly stated to be otherwise.

We now turn to the semantics of normal modal logics. The terminology here follows Scott and Segerberg (Lemmon and Scott [66], Segerberg [71]) though many of the results stem from Kripke [63] and elsewhere.

By a relational frame $\mathcal{F}$ (alternatively called a *model structure*) we understand a pair $\langle W,R \rangle$ in which W is a set and $R \subseteq W^2$. The members of W are sometimes called "possible worlds" (because of the original modal interpretation of the theory in terms of the logic of necessity and possibility) or, more neutrally, *points of reference* or *indices*. By a model $\mathcal{M}$ based on a frame $\mathcal{F}$ we mean a triple $\langle W,R,V \rangle$ where $\mathcal{F} = \langle W,R \rangle$ and V is a function which assigns each $p \in P$ a subset of W. I.e. $V(p) \subseteq W$. By an evaluation $\models$ we understand a three-place relation between a model $\mathcal{M}$, a world $w \in W$ and a wff α of $\mathcal{L}$ satisfying 1.14 - 1.17 below.

We read $\mathfrak{M} \models_w \alpha$ as "α is true at w in model $\mathfrak{M}$" and use $\mathfrak{M} \not\models_w \alpha$ to mean "not $(\mathfrak{M} \models_w \alpha)$".

1.14 If $p \in P$ then $\mathfrak{M} \models_w p$ if and only if $w \in V(p)$

1.15 $\mathfrak{M} \models_w \sim\alpha$ if and only if $\mathfrak{M} \not\models_w \alpha$

1.16 $\mathfrak{M} \models_w \alpha \vee \beta$ if and only if $\mathfrak{M} \models_w \alpha$ or $\models_w \beta$

1.17 $\mathfrak{M} \models_w L\alpha$ if and only if for all w' such that wRw', $\mathfrak{M} \models_{w'} \alpha$

If $\mathfrak{M} \models_w \alpha$ for every $w \in W$ then α is called *valid* in $\mathfrak{M}$. Where $\mathfrak{F}$ is a frame $\langle W,R \rangle$ then α is *valid on* $\mathfrak{F}$ if and only if, for every $\mathfrak{M}$ based on that frame, α is valid in $\mathfrak{M}$. If $\mathfrak{K}$ is a class of frames then α is *valid over* $\mathfrak{K}$ ($\mathfrak{K}$-valid) if and only if α is valid in every frame in $\mathfrak{K}$. (We can also have validity in a class of models.) We say that a model (frame) is a model (frame) *for* or *of* a wff (set of wff) if and only if the wff (all members of the set of wff) is valid in the model (frame).

A class $\mathfrak{K}$ of frames *determines* or *characterises* a logic Λ if and only if for any wff α,

$$\alpha \in \Lambda \Leftrightarrow \alpha \text{ is } \mathfrak{K}\text{-valid.}$$

If α is not valid (in a model, frame, class) we say that it fails (in that model, frame, class).

R is frequently called the *accessibility* relation. Given w, if wRw' then w' is accessible from w. Intuitively the picture is that the necessary is that which not only happens to be true but which must be true, i.e. that which is not only true in the real world but which is true also in all possible worlds, i.e. all worlds accessible from the real world.

We say that a frame $\mathcal{F} = \langle W,R \rangle$ is, respectively, reflexive, transitive, symmetrical, an equivalence frame, if and only if R is a reflexive, transitive, symmetrical, equivalence relation.

THEOREM 1.18

(i) Every frame is a frame for K

(ii) Every reflexive frame is a frame for T

(iii) Every reflexive transitive frame is a frame for S4

(iv) Every reflexive symmetrical frame is a frame for B

(v) Every equivalence frame is a frame for S5

The proof of (i) is that all PC tautologies and $L(p \supset q) \supset (Lp \supset Lq)$ are valid on all frames and that the closure conditions preserve validity on any frame. (Note that uniform substitution does not preserve validity in a model though it does preserve validity on a frame.) The proofs of (ii) - (v) then simply require the establishment of the validity of their particular axioms. This is easily done. (cf. Hughes and Cresswell [68].)

2. *Canonical Models*

In this section we expound a very powerful technique devised by Lemmon and Scott for proving the completeness of many modal logics. The method is based on Henkin's completeness proof for the predicate calculus and proceeds by showing that if Λ is a normal modal logic and $\alpha \notin \Lambda$ then there is a model of Λ in which α fails in some world. This result will follow from the more general result that any Λ-consistent set Γ of wff of $\mathcal{L}$ (a notion to be made precise immediately below) is simultaneously satisfiable in a model of Λ. (I.e. there is a world in the model at which every member of Γ is

We can say that a wff β is a Λ-consequence of a set Γ of wff of $\mathcal{L}$ (written $\Gamma \vdash_\Lambda \beta$) if and only if for some $\alpha_1,\ldots,\alpha_n \in \Gamma$,

$$(\alpha_1 \cdot \ldots \cdot \alpha_n) \supset \beta \in \Lambda$$

Γ is Λ-consistent if and only if not $\Gamma \vdash_\Lambda (p.{\sim}p)$[3]. A set Γ of wff of $\mathcal{L}$ is said to be *maximal* if and only if for any wff α either $\alpha \in \Gamma$ or ${\sim}\alpha \in \Gamma$. A set which is both Λ-consistent and maximal may be called Λ-*maximal consistent* or Λ-*complete*. Obviously a Λ-complete set Γ will contain Λ ; for suppose that $\alpha \in \Lambda$ but $\alpha \notin \Gamma$. Since Γ is maximal ${\sim}\alpha \in \Gamma$, but since Λ is a logic and $\alpha \in \Lambda$ then ${\sim}\alpha \supset (p.{\sim}p) \in \Lambda$ (for Λ contains PC, in particular $\alpha \supset ({\sim}\alpha \supset (p.{\sim}p))$ and MP) and so since ${\sim}\alpha \in \Gamma$ then $\Gamma \vdash_\Lambda p.{\sim}p$ contrary to its assumed consistency.

For a logic Λ the following conditions are equivalent:

2.1 Λ is Λ-consistent

2.2 not both α and ${\sim}\alpha$ are in Λ for any α

2.3 there is some β not in Λ

2.4 for all $p \in P$, $p \notin \Lambda$.

In any of these cases Λ can be spoken of as consistent.

Given any consistent logic Λ we can define what we shall call its *canonical model* $\mathfrak{M}_\Lambda = \langle W_\Lambda, R_\Lambda, V_\Lambda \rangle$ as follows:

2.5 W_Λ is the set of all Λ-complete sets of wff of $\mathcal{L}$

2.6 for any $w, w' \in W_\Lambda$, $wR_\Lambda w'$ if and only if for every wff α of $\mathcal{L}$: if $L\alpha \in w$ then $\alpha \in w'$ (Remember that w and w'

3 An alternative definition of the Λ-consistency of Γ is simply that for no $\alpha_1,\ldots,\alpha_n \in \Gamma$ is ${\sim}(\alpha_1 \cdot \ldots \cdot \alpha_n) \in \Lambda$. We shall sometimes use this characterization.

are sets of wff of $\mathcal{L}$ and so any wff of $\mathcal{L}$ is either in the set or not in it.)

What this principle says is that w' is accessible from w if and only if it has as true all the wff which are necessary in w, and a little reflection ought to show that this reflects what was said about accessibility.

2.7 $V_\Lambda(p) = \{w \in W_\Lambda : p \in w\}$.

THEOREM 2.8

For any wff α *and any* $w \in W_\Lambda$, $\mathfrak{M}_\Lambda \models_w \alpha$ *if and only if* $\alpha \in w$.

This is the crucial theorem of this section and perhaps even deserves to be called the fundamental theorem of modal logic, for from it flow almost all the completeness results so far obtained about modal systems. We need first a few preliminary lemmata.

LEMMA 2.9

If Γ *is* Λ*-consistent then there is some* $w \in W_\Lambda$ *such that* $\Gamma \subseteq w$.

We assume that the wff of $\mathcal{L}$ are given in some enumeration which is to remain fixed throughout the discussion. We define the following sequence of sets.

Γ_0 is Γ.

Γ_{n+1} is $\Gamma_n \cup \{\alpha_n\}$ if this is Λ-consistent (where α_n is the n-th wff of $\mathcal{L}$) and is Γ_n otherwise. Obviously Γ_0 is Λ-consistent and Γ_{n+1} is Λ-consistent if Γ_n is so each of the

Γ_n's is Λ-consistent.

Let $w = \bigcup_{n\geq 0} \Gamma_n$.

Then w is Λ-consistent; for suppose not: then there would be some $\beta_1,\ldots,\beta_k$ such that $\{\beta_1,\ldots,\beta_k\}$ is not Λ-consistent. Since this set is finite there must be a biggest (in the enumeration of wff), say it is α_n. Then $\{\beta_1,\ldots,\beta_k\} \subseteq \Gamma_n$ and so Γ_n would be inconsistent and we have shown that each Γ_n is consistent.

Further w is maximal; for suppose that for some α_n, neither α_n nor $\sim\alpha_n \in w$. Since $\alpha_n \notin w$ then $\Gamma_n \cup \{\alpha_n\}$ is inconsistent (for otherwise Γ_{n+1} would be $\Gamma_n \cup \{\alpha_n\}$ and so $\Gamma_n \cup \{\alpha_n\} \subseteq w$ and so $\alpha_n \in w$) and since $\sim\alpha_n \notin w$ then $\Gamma_m \cup \{\sim\alpha_n\}$ is inconsistent (where $\sim\alpha_n$ is the m-th wff in the enumeration). Now either $m > n$ or $n > m$ and by construction of w either $\Gamma_m \subseteq \Gamma_n$ or $\Gamma_n \subseteq \Gamma_m$. Now $\Gamma_n \vdash_\Lambda \sim\alpha_n$

and $\Gamma_m \vdash_\Lambda \sim\sim\alpha_n$

and so $\Gamma_{\max(n,m)} \vdash_\Lambda \sim\alpha_n$

and $\Gamma_{\max(n,m)} \vdash_\Lambda \sim\sim\alpha_n$

and so $\Gamma_{\max(n,m)} \vdash_\Lambda p.\sim p$

and so $w \vdash_\Lambda p.\sim p$

but w is Λ-consistent.

Whence by reductio ad absurdum w is maximal and thereby Λ-complete which is to say that $w \in W_\Lambda$. Q.E.D.

Given a set Γ of wff of $\mathcal{L}$ we define the set $I(\Gamma)$ as

$$\{\alpha : L\alpha \in \Gamma\}$$

i.e. $I(\Gamma)$ is the set of things which are necessary in Γ.

LEMMA 2.10

If Λ *is a normal modal logic and* Γ *is* Λ*-consistent then for any wff* β *such that* $\sim L\beta \in \Gamma$, $I(\Gamma) \cup \{\sim\beta\}$ *is* Λ*-consistent.*

Proof. Suppose that $I(\Gamma) \cup \{\sim\beta\}$ is not consistent. This means that for some $\alpha_1,\ldots,\alpha_n \in I(\Gamma)$

$$(\alpha_1 . \ldots . \alpha_n) \supset \beta \in \Lambda$$

whence by necessitation (1.12)

$$L((\alpha_1 . \ldots . \alpha_n) \supset \beta) \in \Lambda$$

whence by 1.13 and MP (1.11)

$$L(\alpha_1 . \ldots . \alpha_n) \supset L\beta \in \Lambda$$

whence by L-distribution (from 1.12 and 1.13)

$$(L\alpha_1 . \ldots . L\alpha_n) \supset L\beta \in \Lambda$$

whence $\{L\alpha_1 . \ldots . L\alpha_n, \sim L\beta\}$ is Λ-inconsistent.

But this is a subset of Γ which is supposed to be consistent. Thus by reductio ad absurdum the Lemma holds.

We can now get back to the proof proper of 2.8. The proof is of course by induction on the construction of formulae. The definition assures us that for $p \in P$ and $w \in W_\Lambda$

$$\mathfrak{M}_\Lambda \models_w p \text{ if and only if } p \in w.$$

In the case of the truth functors we rely on the maximal consistency of each w to assure us that $\alpha \in w$ if and only if $\sim\alpha$ is not and that $\alpha \vee \beta \in w$ if and only if either $\alpha \in w$ or $\beta \in w$. This part of the induction follows in the usual way (Hughes and Cresswell [68], p.154f).

The only difficult case is the necessity functor. Our induction hypothesis is that

$$\mathcal{M}_\Lambda \vDash_w \alpha \text{ if and only if } \alpha \in w \quad (\text{for any } w \in W_\Lambda)$$

and we want to show that

$$\mathcal{M}_\Lambda \vDash_w L\alpha \text{ if and only if } L\alpha \in w \quad (\text{for any } w \in W_\Lambda).$$

(1) Suppose $L\alpha \in w$ and wRw'; then by definition of R, $\alpha \in w'$ whence

$$\mathcal{M}_\Lambda \vDash_{w'} \alpha$$

whence (since this is the case for all w' such that wRw')

$$\mathcal{M}_\Lambda \vDash_w L\alpha.$$

(2) Suppose $L\alpha \notin w$

then $\sim L\alpha \in w$ (since w is maximal).

So by Lemma 2.10, since w is consistent, $I(w) \cup \{\sim\alpha\}$ is consistent and so by Lemma 2.9 there is some $w' \in W_\Lambda$ such that $I(w) \cup \{\sim\alpha\} \subseteq w'$.

Now since $I(w) \subseteq w'$, then wRw', and since $\sim\alpha \in w'$ then $\alpha \notin w'$; whence $\mathcal{M} \nvDash_{w'} \alpha$ and, since wRw', $\mathcal{M} \nvDash_w L\alpha$. This proves the inductive step for L and so proves the theorem.

COROLLARY 2.11

Any Λ*-consistent set* Γ *is simultaneously satisfiable in* $\mathfrak{M}_\Lambda$.

Proof. If Γ is Λ-consistent then for some $w \in W_\Lambda$, $\Gamma \subseteq w$ (by Lemma 2.9) whence (by the theorem) for every $\alpha \in \Gamma$

$$\mathfrak{M}_\Lambda \models_w \alpha$$

Q.E.D.

COROLLARY 2.12

A wff α *of* $\mathcal{L}$ *is a member of a logic* Λ *if and only if* α *is valid in* $\mathfrak{M}_\Lambda$.

Proof. Since Λ is contained in every $w \in W_\Lambda$ then every member of Λ will be true in every $w \in W_\Lambda$, i.e. will be valid in $\mathfrak{M}_\Lambda$.

Suppose $\alpha \notin \Lambda$ then $\{\sim\alpha\}$ is Λ-consistent; for if not then $\sim\alpha \supset (p.\sim p) \in \Lambda$, whence, since Λ is a logic, $\alpha \in \Lambda$. Thus for some $w \in W_\Lambda$, $\mathfrak{M}_\Lambda \dashv_w \alpha$, i.e. α is not valid in $\mathfrak{M}_\Lambda$. Since $\mathfrak{M}_\Lambda$ is a model for Λ this shows that a wff is in a logic if and only if it is valid in all models of the logic. (But see below for some remarks on this.) Q.E.D.

COROLLARY 2.13

K *is determined by the class of all frames.*

Proof. We have already shown that every member of K is valid on all frames. Now suppose that $\alpha \notin K$ then, since K is a logic, α is not valid in $\mathfrak{F}_K$ and so not valid on all frames.

Q.E.D.

Now $\mathcal{M}_\Lambda$ will obviously be a model for Λ, since Λ is contained in every $w \in W_\Lambda$, but of course it does not follow that $\mathcal{F}_\Lambda = \langle W_\Lambda, R_\Lambda \rangle$ is a *frame* for Λ. I.e. the fact that a formula is valid with respect to the canonical model on the canonical frame does not entail that it is valid on every model on that frame. A logic which is characterized by a class of frames is called *complete*, and there are normal modal logics which are not complete[4]. A logic which is not complete cannot be valid on its canonical frame. For suppose it were. Then the canonical frame $\mathcal{F}_\Lambda$ would by itself characterize the logic since if $\alpha \in \Lambda$ then α is valid on $\mathcal{F}_\Lambda$ and if $\alpha \notin \Lambda$, α is not valid on $\mathcal{F}_\Lambda$ since α is not valid in $\mathcal{M}_\Lambda$.

The remarks made in the last paragraph show that if we wish to prove the completeness of a normal modal logic Λ it will be sufficient to show that $\mathcal{F}_\Lambda$ is a frame for Λ. In the case of the particular logics we have been discussing this merely means showing that R_Λ satisfies the appropriate condition. E.g. to prove the completeness of T we need merely show that $\mathcal{F}_T$ is reflexive for we know (by theorem 1.18) that any reflexive frame is a frame for T. (This also establishes that T is characterized by the class of all reflexive frames.)

To show that R_T is reflexive it suffices to show that for all wff α and $w \in W_T$, if $L\alpha \in w$ then $\alpha \in w$ and this follows immediately by substitution of α for p in $Lp \supset p$. We can similarly establish that S4 is characterized by the class of all

---

[4] Such logics have been constructed by Fine [74] and Thomason [74]. They are very complicated systems and their only interest seems to be as examples of incomplete logics. A very simple logic which is almost certainly not valid on its canonical frame is $K + LMp \supset MLp$. This logic has been studied by Robert Goldblatt and Kit Fine. It is however complete, although its frames cannot be characterized by any set of first-order conditions on a relation.

reflexive and symmetrical frames and S5 by all equivalence frames. To take the S4 case, suppose $w_1R_\Lambda w_2$ and $w_2R_\Lambda w_3$, then for any wff α if $L\alpha \in w_1$ then $\alpha \in w_2$ and for any wff β if $L\beta \in w_2$ then $\beta \in w_3$. Suppose $L\alpha \in w_1$, then (by $Lp \in LLp$) $LL\alpha \in w_1$, and so $L\alpha \in w_2$, and so $\alpha \in w_3$. I.e. for any α, if $L\alpha \in w_1$, then $\alpha \in w_3$, which is to say $w_1R_\Lambda w_3$. The cases for B and S5 are just as easy.

3. *The Finite Model Property*

If $\alpha \notin \Lambda$ then α fails at some point in $\mathfrak{M}_\Lambda$. Now $\mathfrak{M}_\Lambda$ will not in general be finite, and indeed no interesting modal logic is characterized by a single finite frame[5]. Nevertheless they do have an allied property which is very important. I.e. that for any $\alpha \notin \Lambda$ there is a finite model $\mathfrak{M}$ in which every member of Λ is valid but in which α is not valid. (A finite model is of course just a model in which W is finite.)

Where Λ is a logic in which this is so we say that Λ has the *finite model property*. The finite model property has an obvious connection with decidability, *viz*. that if Λ is axiomatizable and has the finite model property then Λ is decidable. The proof of this is as follows:

Obviously in any finite model $\mathfrak{F}$ it is possible, given any formula, to test effectively whether or not α is valid in that model. Further although there are infinitely many finite models they can be effectively enumerated (up to isomorphism).

This means that if a logic Λ is axiomatizable and has the

[5] This result was proved for the Lewis systems in Dugundji [40]. The methods used are very generally applicable.

finite model property we can set going two effective operations to test whether an arbitrarily presented α is to be in Λ or not.

(A) We generate the members of Λ (we can do this, if Λ is axiomatizable, by generating proofs of its theorems under one of its axiomatizations) and see whether α appears.

(B) We generate the finite models and test α in each of them in turn.

Under the hypothesis α must appear (in a finite time) in the process A or the process B ; if it appears in A then $\alpha \in \Lambda$, if it appears in B then $\alpha \notin \Lambda$. This means that we can decide effectively whether or not $\alpha \in \Lambda$.

Clearly such a procedure will be of no practical use. Practicable decision procedures for many modal systems seem best obtained by the method of semantic diagrams of Hughes and Cresswell [1968], or by some equivalent method.

Notice that it does not follow from the above that a decidable system, even an axiomatizable one, need have the finite model property[6].

(Note that a system with the finite model property must have the finite *frame* property. I.e. if $\alpha \notin \Lambda$ then there is a model $\langle W,R,V\rangle$ on which α fails where $\langle W,R\rangle$ is a *frame* for Λ.)

The most efficient way I know of proving the finite model property is the method of *filtrations*, which is due to Krister Segerberg.

---

[6] An example of a decidable logic without the finite model property is provided in Makinson [69]. Like the incomplete logics mentioned in footnote 4, it is very complicated.

There is an algebraic method used by J.C.C. McKinsey as long ago as 1941 in McKinsey [41] (vide Hughes and Cresswell [68]) but its application to each system requires apparently *ad hoc* (though small) adjustments to the proof. Also in certain cases the method of semantic diagrams found in Hughes and Cresswell [68] shows that certain systems have the finite model property.

The idea of the method of filtration is this. Given a model $\mathfrak{M}$ and a wff α we want to define another model $\mathfrak{M}^*$ which is finite and which falsifies α, at some world, provided $\mathfrak{M}$ does. We do this by identifying in $\mathfrak{M}^*$ all the worlds which do not discriminate in the values they give to subformulae of α. $\mathfrak{M}^*$ is called a *filtration.*

It is convenient to make our precise definitions a little more general. We say that a set Φ of wff of $\mathcal{L}$ is *closed under subformulae* if, and only if, for any wff α and β, if β is a wf part of α then if $\alpha \in \Phi$, $\beta \in \Phi$. Given a model $\mathfrak{M}$ and a set Φ which is closed under subformulae we define an equivalence relation $\approx$ (more strictly $\approx_{\mathfrak{M}/\Phi}$ to indicate its dependence on $\mathfrak{M}$ and Φ) as follows:

for any w and $w' \in W$, $w \approx w'$ if, and only if, for every $\alpha \in \Phi$:

$$\mathfrak{M} \models_w \alpha \leftrightarrow \mathfrak{M} \models_{w'} \alpha .$$

Let W^* be the set of all equivalence classes with respect to $\approx$. (It is easy to see that $\approx$ is an equivalence relation.) This means that any $u \in W^*$ is $\{w' \in W : w \approx w'\}$ for some $w \in W$. We denote u by $[w]$. Let $R^* \subseteq W^{*2}$. We shall say that R^* is *suitable* if, and only if,

3.1 If wRw' then $[w]\ R^*\ [w']$

3.2 If $L\alpha \in \Phi$ and $[w]\ R^*\ [w']$

then $\mathfrak{M} \models_w L\alpha \Rightarrow \mathfrak{M} \models_{w'} \alpha$.

The existence of at least one suitable relation can be guaranteed by defining $[w]\ R^*\ [w']$ to hold if, and only if, for every $L\alpha \in \Phi$, $\mathfrak{M} \models_w L\alpha \Rightarrow \mathfrak{M} \models_{w'} \alpha$. The consistency of this definition requires that if $w_1 \approx w_2$ then $\mathfrak{M} \models_{w_1} L\alpha \Leftrightarrow \mathfrak{M} \models_{w_2} L\alpha$ and if $w_1 \approx w_2$ then $\mathfrak{M} \models_{w_1} \alpha \Leftrightarrow \mathfrak{M} \models_{w_2} \alpha$. But this is guaranteed since both α and $L\alpha$ are in Φ.

For any proposition letter p in Φ we let $[w] \in V^*(p)$ if, and only if, $\mathfrak{M} \models_w p$ (for $p \notin \Phi$ we assume some arbitrary assignment). This definition is consistent for if $[w] = [w']$ then $w \approx w'$ and so, since $p \in \Phi$, $\mathfrak{M} \models_w p \Leftrightarrow \mathfrak{M} \models_{w'} p$.

Where $\mathfrak{M}$ is a model, Φ a set of wff closed under subformulae and W^* and V^* are as described above and R^* is suitable then $\mathfrak{M}^* = \langle W^*, R^*, V^* \rangle$ is called a *filtration of* $\mathfrak{M}$ *through* Φ. We can now prove the fundamental theorem on filtrations.

THEOREM 3.3

If Φ *is closed under subformulae and* $\mathfrak{M}^*$ *is a filtration of* $\mathfrak{M}$ *through* Φ *then for any* $\alpha \in \Phi$ *and any* $w \in W$

$$\mathfrak{M} \models_w \alpha \textit{ if and only if } \mathfrak{M}^* \models_{[w]} \alpha .$$

The proof is by induction on wff of Φ. By definition the theorem holds for propositional letters (provided of course they are in Φ). The induction on the truth functors is pretty obvious but perhaps we should do it for $\sim$ at least to see that it is.

If $\sim\alpha \in \Phi$ then so is α, as Φ is closed under subformulae; so as an induction hypothesis we may assume that

$$\mathfrak{M} \models_w \alpha \text{ if and only if } \mathfrak{M}^* \models_{[w]} \alpha$$

so $\mathfrak{M} \not\models_w \alpha$ if, and only if $\mathfrak{M}^* \not\models_{[w]} \alpha$

and so $\mathfrak{M} \models_w \sim\alpha$ if, and only if, $\mathfrak{M}^* \models_{[w]} \sim\alpha$.

The case with $\vee$ is similar.

The awkward case is of course the modal operator.

Suppose $\mathfrak{M} \not\models_w L\alpha$,

then for some w' such that wRw'

$$\mathfrak{M} \not\models_{w'} \alpha$$

whence by the induction hypothesis

$$\mathfrak{M}^* \not\models_{[w']} \alpha$$

Now since wRw' then $[w]R^*[w']$ (by property 3.1 of R^*) so

$$\mathfrak{M}^* \not\models_{[w]} L\alpha .$$

Suppose $\mathfrak{M}^* \not\models_{[w]} L\alpha$

then for some w' such that $[w]R^*[w']$

$$\mathfrak{M}^* \not\models_{[w']} \alpha, \quad \therefore \text{ by induction hypothesis } \mathfrak{M} \not\models_{w'} \alpha.$$

Now by property 3.2 of R^*, for any $L\beta \in \Phi$ where $[w]R^*[w']$ then if $\mathfrak{M} \models_w L\beta$ then $\mathfrak{M} \models_{w'} \beta$

so since $\mathfrak{M} \not\models_{w'} \alpha$

then $\mathfrak{M} \not\models_w L\alpha$.

This concludes the induction and so proves the theorem.

The fundamental theorem on filtrations mentions nothing about finiteness. This comes in though when Φ is finite. For when Φ is finite the number of distinct assignments to its members will also be finite and so $\mathfrak{M}^*$ will be finite.

Of particular interest is the case where Φ is the set of all subformulae (including α itself) of a wff α. (Call the set Φ_α .)

COROLLARY 3.4

For any wff α *if* α *is not a theorem of* K *then* α *fails in some finite model.*

If α is not a theorem of K then α fails in $\mathfrak{M}_K$ whence, by the fundamental theorem on filtrations, α fails on the model $\mathfrak{M}^*_K$ which is $\mathfrak{M}_K$ filtered through Φ_α.

This corollary might look at first sight as though it gives the finite model property at one blow to all modal logics but a closer look assures us that this is not so.

The finite model property for Λ says that if α is not a theorem of Λ then α fails in a finite Λ-model. We know now that if α is not a theorem of Λ then α fails in $\mathfrak{M}_\Lambda$ and therefore in the filtration $\mathfrak{M}^*$ of $\mathfrak{M}_\Lambda$ through Φ_α. We know that $\mathfrak{M}_\Lambda$ is a model of Λ but we do not know that $\mathfrak{M}^*$ is a model of Λ. In certain cases of course this is trivial.

THEOREM 3.5

K *has the finite model property.*

Since the theorems of K are valid on all models and since $\mathcal{M}^*$ is a model then it is a K-model and so α fails in a finite K-model.

THEOREM 3.6

T *has the finite model property.*

We only need to know that any suitable R^* in a filtration of $\mathcal{M}_T$ is reflexive. From condition 3.1 it follows that since wRw then $[w]R^*[w]$.

THEOREM 3.7

S4 *has the finite model property.*

Here we must define a particular suitable relation. Let $[w]R^*[w']$ if, and only if,

$$\mathcal{M} \models_w L\beta \Rightarrow \mathcal{M} \models_{w'} L\beta$$

for all β such that $L\beta \in \Phi_\alpha$.

We have to show

3.8 that R^* is suitable

3.9 that R^* is transitive.

For 3.8 :

To prove condition 3.1 we note

if wRw' then if $\mathcal{M} \models_w L\beta$, $\mathcal{M} \models_w LL\beta$ and so $\mathcal{M} \models_{w'} L\beta$.

To prove condition 3.2:

if $L\beta \in \Phi$ and $[w]R^*[w']$ then if $\mathfrak{M} \models_w L\beta$, $\mathfrak{M} \models_{w'} L\beta$ and so since S4 contains $L\beta \supset \beta$ then $\mathfrak{M} \models_{w'} \beta$.

For 3.9 R^* is transitive; for suppose

$$[w_1]R^*[w_2] \quad \text{and} \quad [w_2]R^*[w_3]$$

i.e. $\mathfrak{M} \models_{w_1} L\beta \Rightarrow \mathfrak{M} \models_{w_2} L\beta \qquad (\forall\beta : L\beta \in \Phi)$

$\mathfrak{M} \models_{w_2} L\beta \Rightarrow \mathfrak{M} \models_{w_3} L\beta \qquad (\forall\beta : L\beta \in \Phi)$

so obviously $\mathfrak{M} \models_{w_1} L\beta \Rightarrow \mathfrak{M} \models_{w_3} L\beta$

i.e. $[w_1]R^*[w_3]$.

This proof is a little awkward in that it depends on the fact that S4 has $Lp \supset p$, a fact which ought to be irrelevant to establishing transitivity.

Note that we did not define $[w]R^*[w']$ as $\mathfrak{M} \models_w L\beta \Rightarrow \mathfrak{M} \models_{w'} \beta$. This is because with this definition, in order to prove transitivity, we have to make use of $L\beta \supset LL\beta$ and it need not be that $LL\beta \in \Phi_\alpha$. In this case one would have to extend Φ to cover the modalities of subformulae of α. Since there are only finitely many non-equivalent modalities in S4 this would still allow the result, but it may perhaps give an indication of the complexities which can arise.

COROLLARY 3.10

B *has the finite model property.*

In this case we define $[w]R^*[w']$ to hold if, and only if, for any wff β such that $L\beta \in \Phi_\alpha$.

(i) $\mathfrak{M} \models_w L\beta \Rightarrow \mathfrak{M} \models_{w'} \beta$

(ii) $\mathfrak{M} \models_{w'} L\beta \Rightarrow \mathfrak{M} \models_w \beta$.

Obviously R^* is symmetrical. We must show that it is suitable. Suppose wR_Bw'. We know that R_B is symmetrical and hence $w'R_Bw$. Whence both (i) and (ii) hold. Suppose $L\beta \in \Phi_\alpha$. Then if $[w]R^*[w']$, and $\mathfrak{M} \models_w L\beta$, $\mathfrak{M} \models_{w'} \beta$.

THEOREM 3.11

S5 *has the finite model property.*

By defining R^* as for all β such that $L\beta \in \Phi_\alpha$:

$$\mathfrak{M} \models_w L\beta \Leftrightarrow \mathfrak{M} \models_{w'} L\beta$$

we may prove this by analogy with the last two results.

A survey of the relational conditions for a wide range of normal modal logics will be found in Segerberg [71], pp. 47-54 and is beyond the scope of this introductory essay.

BIBLIOGRAPHY

DUGUNDJI, J.

[40] Note on a property of matrices for Lewis and Langford's calculi of propositions, J. Symbolic Logic 5, 1940, 150ff.

FINE, K

[74] An incomplete logic containing S4, Theoria 40, 1974, 23-29.

HUGHES, G.E. and M.J. CRESSWELL

[68] An Introduction to Modal Logic, Methuen, London, 1968.

KRIPKE, S.A.

[63] Semantic analysis of modal logic I, normal propositional calculi, Zeits. f. math. Logik u. Grundl. d. Math. 9, 1963, 67-96.

KRIPKE, S.A.

[65] Semantic analysis of modal logic II, non-normal modal propositional calculi, *The Theory of Models* (ed. J.W. Addison, L. Henkin, A.Tarski), North-Holland, Amsterdam, 1965, 206-220.

LEMMON, E.J. and D.S. SCOTT

[66] Intensional Logic, preliminary draft of initial chapters by E.J. Lemmon, July 1966 (mimeographed).

McKINSEY, J.C.C.

[41] A solution of the decision problem for the Lewis systems S2 and S4 with an application to topology, J. Symbolic Logic 6, 1941, 117-134.

MAKINSON, D.C.

[66] On some completeness theorems in modal logic, Zeits. f. math. Logik u. Grundl. d. Math. 12, 1966, 379-384.

MAKINSON, D.C.

[69] A normal modal calculus between T and S4 without the finite model property, J. Symbolic Logic 34, 1969, 35-38.

SEGERBERG, K.

[68] Decidability of S4.1, Theoria 34, 1968, 7-20.

SEGERBERG, K.

[71] An essay in classical modal logic, Fisosofiska Studier, Uppsala University, 1971.

THOMASON, S.K.

[74] An incompleteness theorem in modal logic. Theoria 40, 1974, 30-34.

* * *

Department of Philosophy, Victoria University of Wellington, Wellington, New Zealand.

A LANGUAGE AND AXIOMS FOR EXPLICIT MATHEMATICS

Solomon Feferman [1]

1. *Introduction*

Systematic explicit mathematics (of various kinds, to be described below) deals with *functions* and *classes* only *via certain means of definition or presentation.* The former operational definitions are called here *rules* or *operations*; definitions of the latter are called *classifications.* In the literature one has also used *constructions* for the first and *predicates, properties, types* or *species* for the second. A new language $\mathcal{L}$ is introduced for which such notions of operation and classification are basic.

Two systems of axioms T_0 and T_1 are formulated in $\mathcal{L}$, the first of which is evident when the operations are interpreted to be given by *rules for mechanical computation.* In T_1 these must be understood instead to be given by *definitions admitting quantification over* N (the natural numbers); T_1 is obtained from T_0 by adjoining a single axiom. In both cases, the classifications may be conceived of as *successively explained or generated* from preceding ones. Some variants and extensions of T_0 and T_1 suggested by the same ideas are also considered.

Several metamathematical results (as to models, conservative extensions, etc.) are obtained for these theories. It is also shown

[1] Research supported by NSF Grant 34091X.

how to formalize directly in them or treat in terms of their models such enterprises as *constructive, recursive, predicative and hyperarithmetic mathematics*. This permits a rather clear view of *what portions of mathematics are accounted for by these systematic redevelopments*.

The following are some distinctive features of the notions axiomatized here, in contrast with current set-theoretical conceptions.

(i) The set-theoretical notions of function and class are viewed *extensionally*, e.g. two classes which have the same members are identical. The notions here are viewed *intensionally*, e.g. two essentially distinct rules may well compute the same values at the same arguments.

(ii) The notions of function and set are *interreducible*: functions may be explained in terms of sets of ordered pairs and sets in terms of characteristic functions. In contrast, the characteristic function associated with a classification A is *not* in general given by a rule. (For example, in the constructive interpretation of T_0, there is no rule for telling which sequences of rationals belong to the classification A of being Cauchy.) There is a *significant asymmetry* in the treatment of the basic notions here. Roughly speaking, rules are taken to be of a quite restricted character, while the properties expressed by classifications may be quite rich. Mathematics consists in discovering which such properties are held by given mathematical objects (e.g. numbers, syntactic expressions, operations and classifications themselves).

(iii) *Self-application* is both possible and reasonable for rules and classifications. The identity operation is given by the rule which associates with any object x the value x. The universal classification V holds of all objects. In general though, *operations are partial*, i.e. have domains which may be a proper part of the universe and so need not be self-applicable. (For example, the operation of differentiation is defined only for certain operations from reals to reals.) Further there may be no extension of a rule f to all of V when there is no test for membership in the domain of f.

(iv) *Operations may be applied to classifications as well as operations.* Important examples are the operation $\underline{c}$ which applies to any A,B to give the *Cartesian product* A × B, and the operation $\underline{e}$ which applied to any A,B gives the *exponentiation* classification B^A holding of just those f which map A into B. Still further we have a *join operation* $\underline{j}$ which applies to any A, f for which fx is a classification B_x whenever x belongs to A; this holds exactly of those pairs z = (x,y) for which x belongs to A and y belongs to B_x. These operations are all guaranteed by the axioms of T_0. In addition, *general principles of inductive generation* in T_0 permit their transfinite iteration.

The classifications generated by $\underline{e}$ applied any finite number of times starting with N are usually called the *finite types*. The objects falling under these classifications are the *functionals of finite type*. The important recognition of this as a constructively admissible notion is due to Gödel [58]. *Constructive theories of transfinite types* have been formulated by Scott [70] and Martin-Löf [prelim.Ms]. The theory T_0 is also constructively justified and is richer than these. Its formulation seems to me to constitute an improvement in other respects as well; however, no detailed comparison is made here[2].

Some ideas for extensions of T_0 are discussed at the conclusion. The interest there is to find much stronger reasonable axioms for classifications; such go beyond current practice if not the needs of explicit mathematics.

2. *The language* $\mathcal{L}$

2.1. *Syntax*

Variables: a,b,c,...,x,y,z

Constants: $\underline{0}$, $\underline{k}$, $\underline{s}$, $\underline{d}$, $\underline{p}$, $\underline{p}_1$, $\underline{p}_2$, $\underline{c}_n(n<\omega)$, $\underline{j}$, $\underline{i}$

Atomic relations: x = y, App(f,x,y), Cl(a), xηa

Atomic formulas: any substitution instance by variables or constants in atomic relations, together with an atomic sentence $\bot$

Connectives and quantifiers: $\wedge$, $\vee$, $\rightarrow$, $\forall$, $\exists$

Formulas are generated from the atomic formulas by the connectives and quantifiers. ϕ, ψ, θ, ... range over formulas.

2.2. *Informal interpretation of the basic syntax*

The variables will be interpreted as ranging over a universe of mathematical objects among which are rules and classifications.

The meaning of the constants will be explained in connection with the axioms.

$\bot$ is a false or absurd proposition.

x = y holds when x and y are identical.

App(f,x,y) holds when f is a rule (or operation) which is defined at the argument x and which has value y when applied

---

[2]Cf. Scott [70] for an extensive discussion of previous work. (Added in proof: cf. the addenda below.)

to x.

Cl(x) holds when x is a classification.

$x\eta a$ holds when x falls under (belongs to, is in) the classification a.

2.3. Abbreviations

$\neg\phi$ for $(\phi \rightarrow \bot)$; $(\phi \leftrightarrow \psi)$ for $(\phi \rightarrow \psi) \wedge (\psi \rightarrow \phi)$;

$\phi(t/x)$ for $\mathrm{Sub}(t,x,\phi)$ – this is also written $\phi(t,\ldots)$ when ϕ is written $\phi(x,\ldots)$;

$\exists! x\phi$ for $\exists x[\phi \wedge \forall y(\phi(y/x) \rightarrow x = y)]$; $x \neq y$ for $\neg(x=y)$;

$\exists x\eta a\phi$ for $\exists x(x\eta a \wedge \phi)$; $\forall x\eta a\phi$ for $\forall x(x\eta a \rightarrow \phi)$.

2.4. Application terms or, simply *terms*, are generated as follows:

(i) Each variable and constant is a term;

(ii) if t_1, t_2 are terms then t_1t_2 is a term.

The informal interpretation is that t_1t_2 is the unique value y of t_1 applied to t_2, if t_1 is defined at t_2. In that case we write $t_1t_2 = y$. Since there may be no y with $\mathrm{App}(t_1,t_2,y)$, strictly speaking terms cannot be considered part of $\mathcal{L}$. Their use with $\mathcal{L}$ can be established by the following *abbreviations* $(t,t_1,\ldots,t_n$ all terms):

$t \simeq y$ for $t = y$, when t is a variable or constant;

$t_1t_2 \simeq y$ for $\exists x_1,x_2[t_1 \simeq x_1 \wedge t_2 \simeq x_2 \wedge \mathrm{App}(x_1,x_2,y)]$;

$(t\downarrow)$ for $\exists y(t \simeq y)$;

$t_1 \simeq t_2$ for $\forall y[t_1 \simeq y \leftrightarrow t_2 \simeq y]$

$\phi(t,\ldots)$ for $\exists y[t \simeq y \wedge \phi(y,\ldots)]$;

$t_1t_2,\ldots,t_n$ for $(\ldots(t_1t_2)\ldots)t_n$.

In each of these abbreviations, the quantified variables on the right are to be distinct and not appear in the expression on the left.

2.5. Classification variables

The classification variables: $A, B, C, \ldots, X, Y, Z$ are introduced by convention to range over the objects for which $Cl(x)$ holds. In other words $\forall X \phi(X)$ is written for $\forall x[Cl(x) \to \phi(x)]$ and $\exists X \phi(X)$ for $\exists x[Cl(x) \wedge \phi(x)]$. Alternatively, we may consider $\mathcal{L}$ as being expanded to a 2-sorted language $\mathcal{L}^{(2)}$ having this new sort of variable which may be used in any of the atomic formulas. We then take as axioms

(1) $$\exists x(x=X),$$

(2) $$Cl(x) \leftrightarrow \exists X(x=X).$$

When we write $\phi(x_1, \ldots, x_n, X_1, \ldots, X_m)$ we are treating ϕ as a formula of $\mathcal{L}^{(2)}$ (all of whose free variables are among $x_1, \ldots, x_n$, $X_1, \ldots, X_m$).

2.6. Elementary formulas

A formula $\phi(x_1, \ldots, x_n, X_1, \ldots, X_m)$ is said to be *elementary* (with respect to classifications) if

(i) its atomic formulas are all of the form $t_1 = t_2$, $App(t_1, t_2, t_3)$ or $t_1 \eta X_i$ for t_1, t_2, t_3 constants or individual variables, and

(ii) ϕ contains no bound classification variables.

Informally, such ϕ does not refer in any way to the general notion of classification. Any given classifications $X_1, \ldots, X_m$ may be tested only with respect to questions as to which objects belong to them.

Each formula ϕ is assigned a Gödel number $\ulcorner\phi\urcorner$ in a standard way. We shall write $\underset{\sim}{c}_{\phi}$ for $\underset{\sim}{c}_{\ulcorner\phi\urcorner}$ when ϕ is elementary.

2.7. Remarks on the choice of language

1. It might be thought that there should also be a predicate $\mathrm{Op}(f)$ expressing that f is an operation. For our purposes this could be introduced instead by definition

$$\mathrm{Op}(f) \leftrightarrow \exists x,y(fx \simeq y),$$

since we never really have to deal with completely undefined rules. However, it would not serve our purposes to define $\mathrm{Cl}(a)$ as $\exists x(x\eta a)$, since it is important to reserve the possibility that a given classification is empty (a common matter for mathematical investigation).

2. Define $A_1 \subseteq A$ by $\forall x[x\eta A_1 \rightarrow x\eta A]$ and $f : A \rightarrow B$ by $\forall x[x\eta A \rightarrow fx\eta B]$. It follows that if $f : A \rightarrow B$ and $A_1 \subseteq A$ then $f : A_1 \rightarrow B$. One might prefer to follow the practice of category theory so that for any f there is at most one A (and at most one B) such that $f : A \rightarrow B$ holds. The syntax and axioms could easily be modified accordingly if desired. For our purposes it is more convenient not to do this. The algebraic notion of morphism can be explained in T_0 in terms of triples (f,A,B) where $f : A \rightarrow B$.

3. The use of the many constants is not essential but is only to simplify statement of the axioms.

3. *The theory* T_0

3.1. Logical axioms are taken to be those of intuitionistic predicate calculus. Use of classical logic (law of excluded middle) will also be permitted when noted explicitly. (If $\mathcal{L}$ is identified with $\mathcal{L}^{(2)}$ then also axioms 2.5(1), (2) are included.)

3.2. The axioms of T_0 are given in five groups I-V. Some further abbreviations are introduced after II and IV.

I (i) $x = y \vee x \neq y$.

(ii) $fx \simeq y_1 \wedge fx \simeq y_2 \to y_1 = y_2$.

(iii) $x \eta a \to Cl(a)$.

II *Basic operations*

(i) (Constant) $\underline{k}xy \simeq x$

(ii) (Substitution) $\underline{s}xy\downarrow \wedge \underline{s}xyz \simeq xz(yz)$

(iii) (Defn. by cases) $(x = y \to \underline{d}abxy \simeq a) \wedge (x \neq y \to \underline{d}abxy \simeq b)$

(iv) (Pairing, projection) $\underline{p}x_1x_2\downarrow \wedge \underline{p}_1z\downarrow \wedge \underline{p}_2z\downarrow \wedge \underline{p}_i(\underline{p}x_1x_2) \simeq x_i$

(v) (Zero) $\neg(\underline{p}xy \simeq \underline{0})$

Abbreviations: (x,y) for $\underline{p}xy$

$(x_1,\ldots,x_{n+1})$ for $((x_1,\ldots,x_n), x_{n+1})$

x' for $(x,\underline{0})$; $\underline{1}$ for $\underline{0}'$.

III *Elementary comprehension scheme.* For each elementary $\phi \equiv \phi(x,y_1,\ldots,y_n, A_1,\ldots,A_m)$:

$$\exists C\{\underline{c}_\phi(y_1,\ldots,y_n,A_1,\ldots,A_m) \simeq C \wedge \forall x[x\eta C \leftrightarrow \phi]\}.$$

IV *Join*

$\forall x\eta A\exists X(fx \simeq X) \to \exists J\{\underline{j}(A,f) \simeq J \wedge \forall z[z\eta J \leftrightarrow \exists x,y(z=(x,y) \wedge x\eta A \wedge y\eta fx)]\}$

V *Inductive generation.* For each formula ψ:

$$\exists I\{\underline{i}(A,R) \simeq I \wedge \forall x\eta A[\forall y((y,x)\eta R \to y\eta I) \to x\eta I]$$

$$\{\forall x\eta A[\forall y((y,x)\eta R \to \psi(y/x)) \to \psi] \to \forall x\eta I.\psi\}.$$

These axioms are fairly transparent. Some fine points of meaning will emerge in the next section. The consistency of T_0 and some direct extensions will be established in §4.[2a]

3.3. Some consequences of the axioms These are treated informally and only sketched.

(1) *Explicit definition.*

By the usual argument for the combinators $\underline{k}$, $\underline{s}$ we can associate with each (application) term t a new term t* such that vars(t*) $\subseteq$ vars(t)-{x} and

$$t^*\downarrow \text{ and } \forall x[t^*x \simeq t].$$

t* is denoted λx.t. Informally, it is reasonable that t* have a value no matter what choice of values for its variables, namely, it is the rule which at x follows out the rule given by t. The special case $\lambda z.\underline{s}xyz$ is incorporated in the axiom for $\underline{s}$, according to which $\underline{s}xy\downarrow$.

(2) *Pairing, n-tupling.* Note that the projection operations $\underline{p}_i$ are defined for all objects. This is in accord with the informal idea that we can tell whether or not an object z is an ordered pair (x_1,x_2) or not. In the first case take $\underline{p}_i z = x_i$, otherwise $\underline{p}_i z = \underline{0}$ (say). Thus $\exists x,y.z = (x,y) \leftrightarrow \underline{p}(\underline{p}_1 z)(\underline{p}_2 z) \simeq z$. From the pairing axiom we derive $(x_1,x_2) = (y_1,y_2) \rightarrow x_1 = y_1 \wedge x_2 = y_2$.

For each $n \geq 2$ and $1 \leq i \leq n$ we can find p_i^n such that $\forall z.p_i^n z\downarrow \wedge \forall x_1,\ldots,x_n.\ p_i^n(x_1,\ldots,x_n) \simeq x_i$. Then given any t we can find t* with vars(t*) $\subseteq$ vars(t) $-\{x_1,\ldots,x_n\}$ such that

$$t^*\downarrow \wedge t^*(x_1,\ldots,x_n) \simeq t.$$

[2a] Myhill has shown that schema III can be replaced by finitely many axioms, in the expected way.

Namely, t^* is $\lambda z.t[p_1^n z/x_1, \ldots, p_n^n z/x_n]$. We write $t^* = \lambda(x_1, \ldots, x_n).t$.

(3) *The Recursion Theorem. (Self-Referential Rules).* The following form

$$\forall f \exists g \forall y_1, \ldots, y_n [g y_1 \ldots y_n \simeq f g y_1 \ldots y_n]$$

is proved essentially as in recursion theory: first define the term $s_1 = \lambda y.zxy$ with vars. z, x. Then for all x, y, z:

$$s_1[z,x]\downarrow \wedge (s_1[z,x])y \simeq zxy.$$

Next, form $\lambda x \lambda y.f(s_1[x,x])y$; this exists, call it h. Then also $g = s_1[h,h]$ exists, and for all y

$$gy \simeq fgy,$$

hence also $g y_1 \ldots y_n \simeq f g y_1 \ldots y_n$.

We can equally well get the theorem in the form

$$\forall f \exists g \forall y_1, \ldots, y_n [g(y_1, \ldots, y_n) \simeq f(g, y_1, \ldots, y_n)].$$

(4) *Elementary operations on classes.* For each elementary $\phi(x, y_1, \ldots, y_n, A_1, \ldots, A_m)$ write

$$\hat{x}\phi(x, y_1, \ldots, y_n, A_1, \ldots, A_m) \text{ for } \underline{c}_\phi(y_1, \ldots, y_n, A_1, \ldots, A_m).$$

We may then make the following *abbreviations:*

V for $\hat{x}.x=x$;

Λ for $\hat{x}.x \neq x$;

$\{a,b\}$ for $\hat{x}(x=a \vee x=b)$;

$\{a\}$ for $\{a,a\}$;

$-A$ for $\hat{x}.\neg(x \eta A)$;

$A \cap B$ for $\hat{x}(x\eta A \wedge x\eta B)$; $A \cup B$ for $\hat{x}(x\eta A \vee x\eta B)$;

$A \times B$ for $\hat{x}(x = (\underline{p}_1 x, \underline{p}_2 x) \wedge \underline{p}_1 x\eta A \wedge \underline{p}_2 x\eta B)$;

$f : A \to B$ for $\forall x(x\eta A \to fx B)$ [i.e. for $\forall x\eta A\, \exists y(fx \simeq y \wedge y\eta B)$];

$$f : A \underset{E_1,E_2}{\to} B \text{ for } (f : A \to B) \wedge \forall x_1\eta A\, \forall x_2\eta A[(x_1,x_2)\eta E_1 \to (fx_1,fx_2)\eta E_2];$$

B^A for $\hat{f}(f:A \to B)$;

$(B,E_2)^{(A,E_1)}$ for $\hat{f}(f : A \underset{E_1,E_2}{\to} B)$;

$\mathfrak{D}(f)$ for $\hat{x}.\exists y(fx \simeq y)$.

(5) *Join and product; union and intersection.* When $\forall x\eta A\, \exists B(fx \simeq B)$ we write B_x for fx and then $\Sigma_{x\eta A} B_x$ for $\underline{j}(A,f)$. Thus $z\eta(\Sigma_{x\eta A} B_x)$ just in case z has the form (x,y) where $x\eta A$, $y\eta B_x$. Under the same hypothesis we can write

$\Pi_{x\eta A} B_x$ for $\hat{g}.\forall x(x\eta A \to gx\eta B_x)$, [i.e. for $\hat{g}.\forall x\{x\eta A \to (x,gx)\eta\underline{j}(A,f)\}$].

If we take $f = \lambda x.B$ then we write $\Sigma_{x\eta A} B$ and $\Pi_{x\eta A} B$ for these, resp. Thus $z\eta(\Sigma_{x\eta A} B) \leftrightarrow z\eta(A \times B)$ and $z\eta(\Pi_{x\eta A} B) \leftrightarrow z\eta(B^A)$.

Again, under the same hypotheses for A, f and B_x we can introduce $\cup_{x\eta A} B_x$ and $\cap_{x\eta A} B_x$ with the usual definitions.

(6) *The natural numbers.* We defined $x' = (x,\underline{0})$. By the axioms for zero and pairing we have:

(i) $x' \neq \underline{0}$

(ii) $x' = y' \to x = y$

(iii) $x = y' \to y = \underline{p}_1 x$.

Thus we may consider $'$ as the successor operation for generating natural numbers and $\underline{p}_1$ the predecessor operation.

These are now used to set up the inductive generation of N. Let $A = \{\underline{0}\} \cup \hat{x}.\exists y\,(x=y') = \{\underline{0}\} \cup \hat{x}\,(x=(\underline{p}_1 x)')$. Let $R = \hat{z}\exists x,y[z=(y,x) \wedge x=y']$. Define $N = \underline{i}(A,R)$. It is seen that

(iii) $\underline{0}\eta N$

(iv) $x\eta N \rightarrow x'\eta N$

(v) for any $\psi(x,\ldots)$:

$$\psi(\underline{0},\ldots) \wedge \forall x[\psi(x,\ldots) \rightarrow \psi(x',\ldots)] \rightarrow \forall x\eta N.\psi(x,\ldots).$$

Using the Recursion Theorem and definition by cases, we find r_N such that for all x, f, a

$$r_N(x,a,f) \simeq \begin{cases} a & \text{if } x = \underline{0} \\ f(\underline{p}_1 x, r_N(\underline{p}_1(x),a,f)) & \text{if } x \neq \underline{0}. \end{cases}$$

In other words

(vi) $r_N(\underline{0},a,f) \simeq a$

$r_N(x',a,f) \simeq f(x,r_N(x,a,f))$.

Hence for any B,

$$r_N : N \times B \times B^{N\times B} \rightarrow B.$$

r_N is a *recursion operator* for N. Using it we may successively define all primitive recursive functions of natural numbers.

The *bounded minimum operator* $(\mu y \leq x)fy \simeq \underline{0}$ and the predicate of *bounded existential quantification* $\exists y \leq x.fy \simeq \underline{0}$ are obtained by primitive recursive defining schemes. Applying the recursion theorem we find g such that

(vii) $$g(f,x) \simeq \begin{cases} (\mu y \leq x) fy \simeq \underline{0} & \text{if } \exists y \leq x . fy \simeq \underline{0} \\ g(f,x') & \text{otherwise.} \end{cases}$$

Let $\mu f \simeq g(f,\underline{0})$; then μf is defined and equal to $\mu x.fx \simeq \underline{0}$ when $\exists x(fx \simeq \underline{0} \wedge \forall y < x.fy\downarrow)$.

Having primitive recursion and μ, we obtain Kleene's enumeration of the partial recursive functions, which associates with each $z \eta N$ a rule $\{z\}$; the total recursive functions are those for which $\{z\} : N \to N$.

Church's Thesis may then be formulated in this language by:

(CT) $$\forall f \eta N^N \, \exists z \eta N \, \forall x \eta N [fx \simeq \{z\}x].$$

(7) *Recursion on inductively generated classifications.* Consider any $\underline{i}(A,R) \simeq I$ in general. By the Recursion Theorem we can find r_I such that

(i) $r_I(x,f) \simeq f(x,r_I)$.

Let $Pd_R(x) = \hat{y}.(y,x)\eta R$. Suppose f is such that

(ii) whenever $x \eta I$ and $g : Pd_R(x) \to V$ then $f(x,g)\downarrow$; then $\forall x \eta I . r_I(x,f)\downarrow$.

r_N is a special case of r_I.

(8) *Tree ordinals.* The countable tree ordinals 0_1 are inductively generated from $\underline{0}$ using successor and N-supremum, where we may identify $\sup_{x \eta N} hx$ with (h,N).

In general, define $\sup_X h$ or $\sup_{x \eta X} hx$ simply as (h,X).

Suppose A consists only of non-empty classifications, i.e. $\forall x \eta A \exists X(x=X \wedge \exists z(z\eta X))$. Thus $z\eta X \wedge X\eta A \leftrightarrow (X,z)\eta J$ where $J = \Sigma_{y\eta A}y$ and $x\eta A \leftrightarrow \exists z.(x,z)\eta J$. We can use the principle of inductive generation to find a classification $I = 0_A$ satisfying the following:

(i) $\underline{0}\eta 0_A$

(ii) $x\eta 0_A \rightarrow x'\eta 0_A$

(iii) $X\eta A \wedge h:(X \rightarrow 0_A) \rightarrow (\sup_X h)\eta 0_A$

(iv) $\psi(\underline{0}) \wedge \forall x[\psi(x) \rightarrow \psi(x')] \wedge \forall X\eta A\forall h[\forall z\eta X\psi(hz) \rightarrow \psi(\sup_X h)] \rightarrow$
$\forall\psi\forall x\eta 0_A.\psi(x)$, for each $\psi(x) = \psi(x,\ldots)$.

Namely 0_A is $\underline{i}(B,R)$ where

$$B = \{\underline{0}\} \cup \hat{x}.\exists y(x=y') \cup \hat{x}.\exists X\eta A\exists h(\forall y\eta X(hy\downarrow) \wedge x = (h,X))$$

and $(y,x)\eta R \leftrightarrow x = y' \vee \exists X,y,z[x = (h,X) \wedge X\eta A \wedge z\eta X \wedge y = hz]$.

When A is empty, 0_A has the same numbers as N. For $A = A_1 = \{N\}$, 0_A is the 0_1 described above. Then for $A = A_2 = \{N,0_1\}$, 0_A is 0_2, etc. We may define A_n and 0_n recursively for $n\eta N$ by: $0_n = 0_{A_n}$ and $A_{n+1} = A_n \cup \{0_n\}$. Then we can pass to transfinite number classes e.g. by taking $A = \cup_{x\eta N}0_x$. More generally, given any C we can associate an 0_x in a natural way with each $x\eta 0_C$, by recursion on 0_C.

(9) *Finite and transfinite types.* Suppose A consists only of non-empty classifications, as in the preceding section. Given a,b,h write $a \dot{\times} b$ for $(\underline{0},a,b)$, $(a \dot{\rightarrow} b)$ for $(\underline{1},a,b)$, $\dot{\Sigma}_{x\eta a}hx$ for $(2,a,h)$ and $\dot{\Pi}_{x\eta a}hx$ for $(3,a,h)$. We inductively generate a classification Typ_A of *A-ary type symbols*, by which is intended that we can form $\dot{\Sigma}$ and $\dot{\Pi}$ over any X in A.

(i) $\underline{0}\eta\mathrm{Typ}_A$

(ii) $a\eta\mathrm{Typ}_A \wedge b\eta\mathrm{Typ}_A \rightarrow (a \dot{\times} b)\eta\mathrm{Typ}_A \wedge (a \dot{\rightarrow} b)\eta\mathrm{Typ}_A$

(iii) $X\eta A \wedge (h{:}X \rightarrow \mathrm{Typ}_A) \rightarrow (\dot{\Sigma}_{x\eta X}hx)\eta\mathrm{Typ}_A \wedge (\dot{\Pi}_{x\eta X}hx)\eta\mathrm{Typ}_A.$

We have in addition a corresponding principle of proof by induction on Typ_A for each formula ψ, which permits definition by recursion on A. Greek letters $\sigma, \tau, \ldots$ are used in the following for type symbols.

When $A = \Lambda$ we call Typ_A the *finite type symbols*. Typ_N is written for $\mathrm{Typ}_{\{N\}}$, the N-ary type symbols. For any A we may define by recursion an operation on Typ_A whose value at each σ is denoted by N_σ, satisfying:

(i) $N_0 = N$

(ii) $N_{\sigma\dot{\times}\tau} = N_\sigma \times N_\tau$, $N_{\sigma\dot{\rightarrow}\tau} = N_\tau^{N_\sigma}$

(iii) $N_{\dot{\Sigma}_{x\eta X}hx} = \Sigma_{x\eta X}N_{hx}$, $N_{\dot{\Pi}_{x\eta X}hx} = \Pi_{x\eta X}N_{hx}$.

It is proved by induction that

(iv) $\sigma\eta\mathrm{Typ}_A \rightarrow \mathrm{Cl}(N_\sigma)$.

We call the operation $\sigma \mapsto N_\sigma$, the *(non-extensional) A-ary type hierarchy*.

An *extensional A-ary type hierarchy* can also be defined. This determines for each $\sigma \in \mathrm{Typ}_A$ two classifications $\overline{N}_\sigma$ and E_σ, defined simultaneously. We write down the clauses for the finite type symbols only:

(i) $\overline{N}_0 = N$, $E_0 = \hat{z}.\exists x(z=(x,x))$.

(ii) $\overline{N}_{\sigma\dot{\times}\tau} = \overline{N}_{\sigma\times\tau}$, $E_{\sigma\dot{\times}\tau} = \hat{z}.\exists x_1,y_1,x_2,y_2[z = ((x_1,y_1),(x_2,y_2)) \wedge (x_1,x_2)\eta E_\sigma \wedge (y_1,y_2)\eta E_\tau]$.

$$\text{(iii)}\quad \overline{N}_{\sigma\dot{\to}\tau} = (\overline{N}_\tau, E_\tau)^{(\overline{N}_\sigma, E_\sigma)} = \hat{f}\{f:\overline{N}_\sigma \to \overline{N}_\tau \wedge \forall x,y\eta\overline{N}_\sigma[(x,y)\eta E_\sigma \to (fx,fy)\eta E_\tau]\}$$

$$E_{\sigma\dot{\to}\tau} = \hat{z}.\exists f,g\{z=(f,g) \wedge f,g\eta\overline{N}_{\sigma\dot{\to}\tau} \wedge \forall x\eta\overline{N}_\sigma (fx,gx)\eta E_\tau\}\ .$$

It is seen that:

(iv) for each finite type symbol σ, $Cl(\overline{N}_\sigma) \wedge Cl(E_\sigma)$ and E_σ is an equivalence relation on $\overline{N}_\sigma$.

It is obvious how to extend the definition of $(\overline{N}_\sigma, E_\sigma)$ to all $\sigma\eta Typ_A$.

Inductively generated classifications of trees with prescribed codings and branchings can be treated in a way similarly to that for the 0_A and Typ_A. Infinite formulas and terms may be considered among such.

3.4. *Non-extensionality of the basic notions*

The classifications $\overline{N}_\sigma$, E_σ introduced in 3.3(9) will be used at various points below to relate certain statements in T_0 to classical mathematical statements concerning extensionally conceived functions and sets. It was stressed in the introduction that for T_0 as a whole the intended conception of the basic notions is intensional. Kreisel has raised the question whether there is an actual *conflict between extensionality and self-application* in this context. He also referred back to a related specific question in Kreisel [71], p.186, as to whether enumeration without repetition conflicts with the axioms for enumerative recursion theory called BRFT in Friedman [71]. As it happens, Friedman stated (loc. cit. p.117) that these are jointly inconsistent. This can be transferred to the present context, since the axioms I(i) and II of T_0 are essentially the same as for an *enumerative system* in Friedman [71], which are in turn equivalent to BRFT. In addition, there is also a conflict of extensionality with self-application for classifications (at least as formulated in T_0). The details are as follows.

The statement of *extensionality for rules* may be considered in either of the following forms:

(i) (a) $\forall f,g[\forall x(fx \simeq gx) \rightarrow f = g]$,

(b) $\forall f,g,h[\forall x(fx \simeq gx) \rightarrow hf \simeq hg]$.

These are equivalent as we see by applying (b) to $h = \lambda x.x$. Similarly the statement of *extensionality for classifications* is considered in the forms:

(ii) (a) $\forall A,B[\forall x(x\eta A \leftrightarrow x\eta B) \rightarrow A = B]$,

(b) $\forall A,B,C[\forall x(x\eta A \leftrightarrow x\eta B) \rightarrow (A\eta C \leftrightarrow B\eta C)]$.

Again these are equivalent by applying (b) to $C = \{A\}$.

Let $Tot(x)$ be the formula $\forall y \exists z(xy \simeq z)$ expressing that x *is a total operation.* Let $e = \lambda x.\lambda y.\underline{d}\underline{0}\underline{0}(xy)\underline{0}$. By 3.3(1) we can prove in T_0 that for all x,y: $ex\downarrow$ and $exy \simeq \underline{d}\underline{0}\underline{0}(xy)\underline{0}$. Thus

(iii) $Tot(e)$, $\forall y(exy\downarrow \leftrightarrow xy\downarrow)$, $\forall y(xy\downarrow \rightarrow exy \simeq \underline{0})$ and $Tot(x) \leftrightarrow Tot(ex)$.

Let $0^* = \lambda x.\underline{0}$. If extensionality held for rules we would have

(iv) $Tot(x) \leftrightarrow ex \simeq 0^*$.

Put $n = \lambda \underline{z}.\underline{d}\underline{1}\underline{0}\underline{z}\underline{0}$ so

(v) $Tot(n)$ and $\forall z \neg(nz \simeq z)$.

Finally, let $f = \lambda x.\underline{d}(n(xx))\underline{0}\ (ex)0^*$ so that for all x:

$$\text{(vi)} \quad fx \simeq \begin{cases} n(xx) & \text{if } ex \simeq 0^* \\ \underline{0} & \text{otherwise.} \end{cases}$$

$\mathrm{Tot}(f)$ because $\mathrm{Tot}(e)$ and $\mathrm{Tot}(x)$ whenever $ex \simeq 0^*$; hence $ef \simeq 0^*$ and $ff \simeq n(ff)$ which is impossible by (v). It is seen by this argument that *extensionality for rules is inconsistent with axioms* I(i), (ii) *and* II *of* T_0.

Turning to classifications, let $\phi(y,x)$ be $\exists z.xy \simeq z$ and $c = \underline{c}_\phi$. By Axiom III, $cx\downarrow$ for all x, and $\mathrm{Cl}(cx)$. Thus

(vii) $\mathrm{Tot}(c)$ and $\forall y[y\eta(cx) \leftrightarrow \exists z.xy \simeq z]$, so $\forall y(y\eta(cx)) \leftrightarrow \mathrm{Tot}(x)$.

If extensionality held for classifications we would have

(viii) $\mathrm{Tot}(x) \leftrightarrow cx \simeq V$.

In the definition (vi) of f above replace e by c and 0^* by V. It is seen that *extensionality for classifications is inconsistent with axioms* I, II, III *of* T_0.

4. *Metamathematical results concerning* T_0 *and related theories*

4.1. A recursion-theoretic model of T_0. I believe that the general informal interpretation given in 2.2 should be clear enough for one to recognize that the axioms of T_0 are correct, hence consistent. As a particular informal interpretation, V may be taken to consist of all expressions generated from finitely many (> 1) symbols, and $fx \simeq y$ to hold whenever f is a program (represented in V) for a mechanical computation which yields the value y at the argument x. Further, the classifications are taken to be certain finite or infinite formulas ϕ represented in V, successively built up and with meaning explained according to the axioms III-V; $a\eta\phi$ is written when ϕ holds of a. Evidently, (CT) is also correct in this interpretation.

The following serves to establish the consistency of T_0 + (CT) assuming set-theory and classical logic. (It is not excluded that one may accept both this and the preceding.) Here "model" is used in its usual sense so that also the laws of classical logic may be applied in T_0. The proof itself gives set-theoretical form to the informal interpretation just given.

THEOREM 4.1.1

There is a model of T_0 *in which the range of the variables is the set* ω *of natural numbers and* $fx \simeq y$ *is interpreted as* $\{f\}(x) \simeq y$. (CT) *is true in this model.*

Note: We are using ω for the natural numbers, to distinguish it from N which is to be interpreted as a particular element of ω. $\{f\}$ $(f=0,1,2,\ldots)$ is a standard enumeration of the partial recursive functions on ω. Church's Thesis (CT) is formulated in terms of N as in 3.3(6).

The proof is straightforward and will only be sketched. By ordinary recursion theory we may choose numbers k, s, d, p, p_1, p_2 so that the basic operation axioms II are satisfied. (Pairing is chosen so that $(x,y) \neq 0$.) Take $\{c_n\}z \simeq (1,n,z)$, $\{j\}(a,f) \simeq (2,a,f)$, $\{i\}(a,r) \simeq (3,a,r)$.

A set Cl_α is defined by transfinite recursion for each ordinal α; the predicate $a \in \cup_\alpha Cl_\alpha$ will be the interpretation of $Cl(a)$. We also define $\{x : x \in \omega \ \& \ x\eta a\}$ for $a \in Cl_\alpha$ along with Cl_α. Let Cl_0 be empty. Suppose given Cl_α and η restricted to $\omega \times Cl_\alpha$. Then for elementary $\phi(x,y_1,\ldots,y_n,A_1,\ldots,A_m)$, the truth of ϕ as a function is well-determined when we assign to each A_i the value a_i in Cl_α. $Cl_{\alpha+1}$ consists of all numbers b obtained by one of the following clauses (i)-(iv); $\{x : x\eta b\}$ is also defined for each of these:

(i) $b \in Cl_\alpha \Rightarrow b \in Cl_{\alpha+1}$; $x\eta b$ is unchanged.

(ii) If $\phi(x,y_1,\ldots,y_n,A_1,\ldots,A_m)$ is elementary and $a_1,\ldots,a_m \in Cl_\alpha$ and $b = \{c_\phi\}(y_1,\ldots,y_n,a_1,\ldots,a_m)$ then $b \in Cl_{\alpha+1}$;

$$x\eta b \Leftrightarrow \phi(x,y_1,\ldots,y_n,a_1,\ldots,a_m) \text{ is true.}$$

(iii) If $a \in Cl_\alpha$ and $\forall x\eta a[\{f\}(x) \in Cl_\alpha]$ then $b = \{j\}(a,f) \in Cl_{\alpha+1}$;

$$z\eta b \Leftrightarrow \exists x,y[z = (x,y) \wedge y\eta\{f\}(x)].$$

(iv) If $a, r \in Cl_\alpha$ then $b = \{i\}(a,r) \in Cl_{\alpha+1}$;

$$u\eta b \Leftrightarrow u \in \cap X\{X \subseteq \omega \wedge \forall x\eta a[\forall y((y,x)\eta r \Rightarrow y \in X) \Rightarrow x \in X]\}.$$

If α is a limit number take $Cl_\alpha = \cup_{\beta<\alpha} Cl_\alpha$. It is then seen that $Cl = \cup Cl_\alpha[\alpha \text{ countable}]$ provides a model of the axioms III-V. In this model, the interpretation of N (a particular $\{i\}(a,r)$) is such that

$$(iv)^N \qquad u\eta N \Leftrightarrow u \in \cap X\{0 \in X \wedge \forall x(x \in X \Rightarrow (x,0) \in X)\}.$$

Thus N is isomorphic with ω, with the successor operation on ω

corresponding to the operation $x \mapsto (x,0)$ which is x' in $\mathcal{L}$. This recursive isomorphism is used to show that (CT) is true in the model.

(1) *Remarks.* (a) Except for (CT) the same method of proof may be applied to give a model of T_0 on any structure $(M,App,0,k,s,d,p,p_1,p_2)$ which satisfies the axioms I(ii) and II of T_0. As mentioned in §3.4 above these are essentially the same as the enumerative systems of Friedman [71] (§1). Recursion theory on admissible sets satisfying Σ_1-uniformization gives a wide and familiar class of examples of such structures. The theory of prime computable functions on any structure (Moschovakis [69]) provides still further examples; this simplifies when the structure is on a transitive set closed under pairing.

(b) It might be thought we could just as well get a model with $Cl = \omega$, e.g. simply by taking $x \not\eta a$ for each $a \notin \cup_\alpha Cl_\alpha$ as defined in the proof. However, $T_0 + \forall a Cl(a)$ is inconsistent: for it follows from $\forall a Cl(a)$ that there is a classification $B = \Sigma_{a \eta V} a$; then take $x \eta C \leftrightarrow (x,x) \not\eta B$.

(c) The proof of Theorem 4.1.1 can be formalized in classical 2nd order analysis, by taking Cl and the graph of the characteristic function of η as the least pair of sets satisfying certain (arithmetical) closure conditions[3]. It may be of interest to see if there are some familiar subsystems of analysis (or set theory) which are of the same strength as T_0.

---

[3]To be more precise, T_0 + (CT) may be translated into the subsystem $(\Pi^1_2\text{-CA})$ + (BI) of 2nd order analysis.

(2) *Question. Is* $T_0^{(c)}$ *(in classical logic) no stronger than* $T_0^{(i)}$ *(in intuitionistic logic)?*

The usual reduction by Gödel's $\neg\neg$-translation breaks down with iterated inductive definitions. (cf. Zucker [73] for some of the problems involved in comparing classical with intuitionistic theories of such definitions.)

(3) *Explicit definability and disjunction properties.* These have been established for most intuitionistic systems which have been considered in the literature, including various theories of species and sets (cf. (Troelstra [73] for much of this and further references). Contrary to my expectation, Myhill pointed out that they fail for T_0 for the simple reason that though $(a = b \vee a \neq b)$ is an axiom, the theory does not decide which of $(t_1 = t_2)$ or $\neg(t_1 = t_2)$ holds for various closed (and defined) t_1, t_2. He conjectures that the properties in question do hold for some simple extensions of T_0 by such basic sentences. In any case, as Kreisel and Troelstra have both emphasized, the fact that a theory enjoys these properties is neither necessary nor sufficient for its constructivity (cf. Troelstra [73], p.91).

Let FT be the language of finite type theory with induction and recursion over N, e.g. the language $N - HA^{\omega}$ of (Troelstra [73]) I.6, expanded to include product types. We have variables $x^{\sigma}, y^{\sigma}, \ldots$ and an equality relation $=_{\sigma}$ for each type symbol σ. If θ is a sentence of $\mathcal{L}_{FT}$ let $\theta^{(\overline{N}_{\sigma}, E_{\sigma})}$ or simply $\theta^{(\overline{N}, E)}$ be its translation into $\mathcal{L}$, taking the variables of type σ to range over the members of $\overline{N}_{\sigma}$ and translating $x =_{\sigma} y$ by $(x,y)\eta E_{\sigma}$.

Note that the interpretation of $(\overline{N}_\sigma, E_\sigma)_\sigma$ in the model $(\omega, Cl, \ldots)$ of Theorem 1 is in 1-1 correspondence with the *hierarchy of hereditarily extensional (recursive) operations* $(HEO_\sigma, \equiv_\sigma)_\sigma$ (due to Kreisel; cf. (Troelstra [73], II.4). This is an isomorphism with respect to pairing, projections and application. $\theta^{(HEO,\equiv)}$ is written for the interpretation of θ in this hierarchy.

COROLLARY 4.1.2

If θ *is a sentence of* $\mathcal{L}_{FT}$ *and* $T_0 \vdash \theta^{(\overline{N},E)}$ *(with classical logic) then* $\theta^{(HEO,\equiv)}$ *is true.*

The definition of $(HEO_\sigma, \equiv_\sigma)$ can be extended in an obvious way to transfinite types, for which the corollary continues to hold.

4.2. Set-theoretical interpretation of T_0. Here we want an interpretation which matches up the extensional finite type hierarchy $(\overline{N}_\sigma, E_\sigma)_\sigma$ with the set-theoretical *maximal type structure* $(M_\sigma)_\sigma$ defined by: $M_0 = N$, $M_{\sigma \dot{\times} \tau} = M_\sigma \times M_\tau$ and

$$M_{\sigma \dot{\to} \tau} = \{F \mid Fun(F) \wedge Dom(F) = M_\sigma \wedge F : M_\sigma \to M_\tau\}.$$

These are defined in Zermelo set-theory (ZS).

To extend the following group of results to transfinite types we need Zermelo-Fraenkel set theory; thus they are stated for ZF instead of ZS.

THEOREM 4.2.1

(i) For any model $\mathfrak{U} = (A, \in)$ *of* ZF *we can associate a model* $\mathfrak{U}^*$ *of* T_0 *in which* $V = A$.

(ii) The interpretation of the $\overline{N}_\sigma, E_\sigma$ *in* $\mathfrak{U}^*$ *is such that* $\overline{N}_\sigma/E_\sigma$ *is in 1-1 correspondence with* M_σ *of* $\mathfrak{U}$ *for each finite type* σ. *This correspondence is the identity on* N *and preserves pairing, projections and applications.*

The idea of the proof is to use the theory of prime computable functions (Moschovakis [69]) for the structure $\mathfrak{A}' = (A, \in, F, \langle a \rangle_{a \in A})$, where $F(u,x)\ u(x)$; this gives an enumerative system of functions $PR(\mathfrak{A}')$ which includes every constant function[3a]. Take $Cl_0 = \{0\} \times A$; for $x, y \in A$ take $x\eta(0,y) \leftrightarrow \exists z[x=(0,z) \wedge z \in y]$. Then proceed to determine Cl_α and η on Cl_α for $\alpha > 0$ just as in 4.1.1. The definition of Cl_0 gives an injection of $(A, \in)$ in (Cl, η). The correspondence is set up recursively. For example, associate with each f in $\overline{N}_{0 \dot{\rightarrow} 0}$ the function $\lambda x \in N.fx$ in $M_{0 \dot{\rightarrow} 0}$; this association is surjective since every element of $M_{0 \dot{\rightarrow} 0}$ is a partial function in $PR(\mathfrak{A}')$ and equivalent members of $N_{0 \dot{\rightarrow} 0}$ correspond to the same function.

For θ in L_{FT} let $\theta^{(M)}$ be its interpretation in the maximal type structure, taking $=_\sigma$ to be $=$ for each σ.

COROLLARY 4.2.2

If θ *is a sentence of* $\mathcal{L}_{FT}$ *and* $T_0 \vdash \theta^{(\overline{N},E)}$ *then* $ZF \vdash \theta^{(M)}$.

4.3. Realizing axioms of choice. By the *relative axiom of choice schema* in $\mathcal{L}$ we mean all formulas:

[3a] J. Stavi pointed out to me that my previous formulation of this argument in terms of admissible sets worked only for ZFC and then only with some additional considerations. He suggested the use of Moschovakis [69] instead.

(AC) $\forall x \eta A\ \exists y\ \phi(x,y) \rightarrow \exists f\ \forall x \eta A\ \phi(x,fx)$.

For particular A we denote this by (AC_A). This may be analyzed as a consequence of

(AC_V) $\forall x\ \exists y\ \phi(x,y) \rightarrow \exists f\ \forall x\ \phi(x,fx)$

and a principle called Independence of Premiss:

(IP) $\forall x \eta A\ \exists y\ \phi(x,y) \rightarrow \forall x\ \exists y(x \eta A \rightarrow \phi(x,y))$.

It will be shown here that (AC) gives a conservative extension of a certain subtheory $T_0^{(-)}$ of T_0, where the use of the existential quantifier in defining properties is restricted to the cases in which that use can be made explicit. It can be shown that T_0 itself is consistent with some instances of (AC), including (AC_V).

The axiom groups I, II for $T_0^{(-)}$ are the same as for T_0. III-V are modified as follows:

III'. Elementary comprehension schema is restricted to ϕ which do not contain existential quantifiers.

III''. We add axioms for operations $\underline{e}$, $\underline{dm}$ where $\underline{e}(A,B) \simeq B^A$, $\underline{dm}f \simeq \mathcal{D}(f)$.

IV is as before for join; to this is added an axiom

IV' for product, $\underline{pr}(A,f) \simeq \Pi_{x\eta A} fx$ under the same hypothesis.

V. Inductive generation is modified to an axiom for an operation $\underline{i}^*(A,S)$, replacing '$(y,x)\eta R$' throughout by '$\exists z.(y,x,z)\eta S$'. Again the logic of $T_0^{(-)}$ is taken to be intuitionistic, unless otherwise noted.

It is seen that $T_0^{(-)}$ has practically the same mathematical consequences as those indicated in 3.3 for T_0. Continuing the idea here we could consider a theory $T_0^{(--)}$ in which also the use of disjunction in defining properties is restricted. The only loss then are the general $\cup$ and $\bigcup$ operations; the disjoint union always serves for the remaining mathematical uses.

The classes $\mathcal{F}_0, \mathcal{F}_1$ of formulas of $\mathcal{L}$ are defined as follows.

(i) each $\mathcal{F}_i$ contains all atomic formulas and is closed under the operations of $\wedge$, $\vee$ and universal quantification;

(ii) If ϕ is in $\mathcal{F}_0$ and ψ is in $\mathcal{F}_1$ then $(\psi \to \phi)$ is in $\mathcal{F}_0$ and $(\phi \to \psi)$ is in $\mathcal{F}_1$;

(iii) If ϕ is in $\mathcal{F}_0$ then $\exists x \phi$ is in $\mathcal{F}_0$.

Thus all formulas without $\exists$ are in both $\mathcal{F}_0$ and $\mathcal{F}_1$.

THEOREM 4.3.1

$T_0^{(-)}$ + (AC) *is a conservative extension of* $T_0^{(-)}$ *for formulas in* $\mathcal{F}_0$; *in fact, if* $T_0^{(-)} + (AC) \vdash \exists x.\phi(x)$ *where* ϕ *is in* $\mathcal{F}_0$ *then* $T_0^{(-)} \vdash \phi(t)$ *for some application term* t.

Again the proof is sketched. We associate with each formula ϕ a formula ρ_ϕ with one new free variable f which we write $f\rho\phi$ and read "f realizes ϕ".[4]

(i) for ϕ atomic, $f\rho\phi$ is $(f=f) \wedge \phi$;

[4] Cf. Troelstra [73], Ch. III for similar variants of Kleene's definitions of realizability.

(ii) $f\rho(\phi \wedge \psi)$ is $(\underline{p}_1 f)\rho\phi \wedge (\underline{p}_2 f)\rho\psi$;

(iii) $f\rho(\phi \vee \psi)$ is $(\underline{p}_1 f)\rho\phi \vee (\underline{p}_2 f)\rho\psi$;

(iv) $f\rho(\phi \to \psi)$ is $\forall g[g\rho\phi \to fg\rho\psi]$;

(v) $(f\rho\forall x\phi)$ is $\forall x(fx\rho\phi)$

(vi) $(f\rho\exists x\phi)$ is $\exists x[f = (\underline{p}_1 f,x) \wedge (\underline{p}_1 f)\rho\phi]$.

Also with each ϕ is associated in $T_0^{(-)}$ a non-empty class $Typ(\phi)$ which includes all rules f which may realize ϕ. In particular, we take $Typ(\psi \to \phi) = Typ(\phi)^{Typ(\psi)}$.

The following may be shown:

(vii) If $T_0^{(-)} + (AC) \vdash \phi$ then $T_0^{(-)} \vdash (t\rho\phi)$ for some application term t.

(viii) If $\phi \in \mathcal{F}_0$ then $T_0^{(-)} \vdash \exists f(f\rho\phi) \to \phi$ and if $\psi \in \mathcal{F}_1$ then $T_0^{(-)} \vdash \psi \to \forall f \eta Typ(\psi)(f\rho\psi)$.

The theorem follows directly from (vii), (viii).

COROLLARY 4.3.2

$T_0^{(-)} + (AC) + (CT)$ *is consistent.*

More generally, $T_0^{(-)} + (AC)$ is consistent with any ψ such that $\exists f(f\rho\psi)$ is true in the model of 4.1.1.

A similar theorem can be established for $T_0^{(--)}$, by appropriately modifying the definition of $\mathcal{F}_0$, $\mathcal{F}_1$. One can also obtain analogous results for T_0 in place of $T_0^{(-)}$, but only for certain extensions (AC_A) of T_0 – roughly speaking for those classes A whose existence is established in $T_0^{(-)}$. But this seems to require a somewhat more delicate treatment of realizability starting with $f\rho(x\eta A)$ written as $(f,x)\eta A^*$ (A^* a variable associated with A).

It is easily seen that $T_0^{(-)} + (AC) + (CT)$ is inconsistent with classical logic.

QUESTION[5]

Is $T_0 + (AC_V) \pm (CT)$ *consistent with classical logic?*

Using the primitive recursive relation $<$ on N, the schema for the *least element principle* is the following:

(LE) $$\exists x\eta N.\phi \rightarrow \exists x\eta N[\phi \wedge \forall y(y < x \rightarrow \neg\phi(y/x)].$$

There is a corresponding rule (LER), to infer the conclusion of this implication from the hypothesis. If that were a derived rule of $T_0^{(-)} + (AC)$ then whenever $T_0^{(-)} + AC \vdash \exists x\eta N.\phi$, the conclusion ψ would be proved to be realizable in $T_0^{(-)}$. But then $\exists f(f\rho\psi)$ would be true in the model of 4.1.1. It would follow that if ψ is a number-theoretical statement then ψ is *recursively realizable.* Hence by the result of Kleene (Kleene [52], p.511) we obtain:

COROLLARY 4.3.3

(LER) *is not a derived rule of* $T_0^{(-)} + (AC)$ *even for hypotheses provable in* $T_0^{(-)}$.

5. *Relations with constructive and recursive mathematics*

The discussion in this section will be sketchy and programmatic.

---

[5]Raised by R. Statman.

5.1. *Constructivity* is understood here in the sense of *intuitionism*[6]. Bishop [67], [70] takes a more restrictive position but within which he redevelops substantial portions of mathematics (cf. also Bishop, Chang [72]). The essential difference is that he rejects use of Brouwer's notion of *choice sequence*, using alternative means for the treatment of analysis and topology. The dispensability of choice sequences was theoretically justified in some systems of intuitionistic analysis by Kreisel, Troelstra [70].

$\mathcal{L}$ is informally interpreted in intuitionistic terms as follows: $fx \simeq y$ holds if f is a *construction* (or *constructive function*) which gives the value y when applied to x. The notion of classification is interpreted as that of *species* (or *type*) and $x\eta A$ by: x belongs to the species A. Bishop's notion of *set* may be identified more particularly with pairs (A,E) for which A, E are classifications and E is an equivalence relation on A. (Inversely, classifications may be explained in Bishop's terms as sets equipped with the relation of literal identity; for alternative explanations cf §7.3 below.)

There is no notion in $\mathcal{L}$ which expresses that of choice sequence. Nor is there a means of expressing in $\mathcal{L}$ the notion of *constructive proof*. The latter is essential for the intuitionistic reduction of logic to mathematics (cf. Kreisel [65], §2).

(1) *Claim.* T_0 *is constructively correct.*

It seems to me that this should be accepted under all the explanations

[6] cf. e.g. Heyting [72], Kreisel [65], §2, Tait [68], and Troelstra [69] for various explanations of this position.

of the constructivist position mentioned here; cf. particularly the line of argument in (Tait [68]).

(2) *Claim. All of Bishop's work* (Bishop [67], Bishop, Chang [72]) *can be formalized in* T_0.

A related claim for a portion of Bishop [67], using a somewhat weaker theory of finite types in place of T_0, has been made by Goodman and Myhill [72]. However, they did not see how to deal with Bishop's general concept of set. As explained above, this is handled directly in T_0. Actually for (2) one should need only that part of V (along with I-IV) required to obtain N and inductively generated N-branching trees (used for countable ordinals and Borel sets).

5.2. *Relations with recursive mathematics*

There have been a number of investigations of *recursive analogues* of classical notions[7]. These yield results concerning statements $\phi^{(rec)}$ formulated in recursive terms analogous to some classical statement $\phi^{(cl)}$. The results for which $\phi^{(rec)}$ is true are often called *positive* while those for which it is false are called *negative*. For example, the theorem on the existence of the maximum of a continuous function on a closed interval has a positive recursive analogue; the statement that the maximum is taken on at some point has a negative analogue.

The interest of such a program obviously depends to a good extent on the choice of $\phi^{(rec)}$ given $\phi^{(cl)}$. It may be asked whether reasonable requirements for this choice can be formulated in

[7] For set theory cf. e.g. Dekker, Myhill [60], Crossley [69]; for algebra, Mal'cev [71], Rabin [62], Ershov [68]; for analysis, Specker [59], the Markov school (Sanin [68] and Tseytin, Zaslavsky, Shanin [66]); and for topology, Lacombe [59].

precise terms. The following is an example of such for a class of statements that covers many of the actual examples.

(1) *Suppose* $\phi^{(cl)}$ *is provably equivalent in set theory* (say ZFC) *to* $\theta^{(M)}$ *where* θ *is a sentence of* $\mathcal{L}_{FT}$; *then* $\theta^{(HEO,\equiv)}$ *is a candidate as the choice for* $\phi^{(rec)}$. (For example, where $\phi^{(cl)}$ concerns real numbers, θ will deal instead with Cauchy sequences of rationals under an equivalence relation.) When a choice is made according to (1) we can hope to learn much more from a positive result, in the light of 4.1.2 and 4.2.2.

(2) *Conjecture. For each known positive result of recursive mathematics of the form* $\theta^{(HEO,\equiv)}$ *where* θ *is a sentence of* $\mathcal{L}_{FT}$ *we have* $T_0 \vdash \theta^{(\overline{N},E)}$. We may regard $(\overline{N},E)$ in this case as a *constructive analogue* (or *substitute*) $\phi^{(cv)}$ of $\phi^{(cl)}$ which in fact is a *generalization* of both $\phi^{(rec)}$ and $\phi^{(cl)}$. Indeed, by 4.2.1(ii), read classically $\phi^{(cv)}$ *is equivalent* to $\phi^{(cl)}$. These relationships illustrate the following.

(3) *General expectation.* (i) *Each classical theorem* $\phi^{(cl)}$ *for which a recursive analogue* $\phi^{(rec)}$ *has been considered has a constructively meaningful form* $\phi^{(cv)}$. (ii) *When* $\phi^{(rec)}$ *is true,* $\phi^{(cv)}$ *is constructively provable.*

We may add, for the particular language and axioms considered:

(3) (iii) *When* $\phi^{(rec)}$ *is false then* $\phi^{(cv)}$ *is independent of* T_0.

Obviously we can also get independence results for any T such that $\phi^{(cv)}$ is interpreted as $\phi^{(rec)}$ in a suitable model.

REMARKS

(a) Requirements of the kind (1) above are only a first step to finding appropriate recursive and constructive analogues of classical statements. For even if such a choice is made, we may have θ_1, θ_2 with $\phi^{(cl)}$ equivalent in set theory to both $\theta_1^{(M)}$ and $\theta_2^{(M)}$, yet θ_1 is true and θ_2 is false in $(HEO,\equiv)$. For example, the classical theorem may have the form $\exists f\ \forall x{\in}A\ \phi(x,f(x))$. If A is definable in the form $\exists y.(x,y) \in B$, it may not be possible to find f as an effective function of x alone. On the other hand, for the classically equivalent statement $\exists g\ \forall(x,y) \in B\phi(x,g(x,y))$ we may be able to find g as an effective function. This is a well-recognized technique for finding positive recursive or constructive substitutes of classical theorems. For a smooth-running positive development one usually makes a choice of notions (e.g. Cauchy sequences considered only as paired with a rate-of-convergence function) which automatically involve this technique wherever needed.

(b) It is possible that the theory $T_0^{(-)}$ + (AC) could lend itself to the purpose of (3) above in the following way. First find a statement ψ of $\mathcal{L}$ which is equivalent in set theory to $\phi^{(cl)}$ and such that $T_0^{(-)} + (AC) \vdash \psi$. Then take $\phi^{(cv)}$ to be $\exists f(f\rho\psi)$. Note that $\phi^{(cv)}$ is also classically equivalent to $\phi^{(cl)}$ but now $T_0^{(-)} \vdash \phi^{(cv)}$. Finally, let $\phi^{(rec)}$ be the interpretation of $\phi^{(cv)}$ in the recursion-theoretic model of 4.1.1. Even with this approach one would still have to go through some of the work of the preceding remark, since when applying (AC_A) we can only use classifications proved to exist in $T_0^{(-)}$ + AC; these in general do not include existentially definable A.

(c) Since T_0 is not extensional, when dealing with

generalizations of classical theorems it is necessary throughout to replace sets by pairs (A,E) where E is an equivalence relation on A. Similarly, instead of algebraic structures $\mathfrak{A} = (A ; R_1,\ldots,f_1,\ldots,a_1,\ldots)$ one will consider more generally pairs $(\mathfrak{A},E)$ of such for which E is a congruence relation on $\mathfrak{A}$. The operation $\mathfrak{A} \rightarrow \mathfrak{A}/E$ cannot be performed, but when E is carried along, this is not necessary.

Call a classification X *decidable relative to* B if $X \subseteq B$ and X has a characteristic function g relative to B, i.e.

$$\forall x \eta B[(g_x \simeq \underline{0} \vee g_x \simeq \underline{1}) \wedge (g_x \simeq \underline{0} \leftrightarrow x \eta X)].$$

Call A *denumerable* if there exists $h : N \underset{\text{onto}}{\rightarrow} A$. When A is denumerable and E is a decidable congruence relation (relative to A^2) we can choose representatives of the E-equivalence classes and form a structure $\mathfrak{A}/E$. If each relation of $\mathfrak{A}$ is also decidable (relative to the appropriate A^n) then in the model of 4.1.1, $\mathfrak{A}/E$ is isomorphic to a *recursively enumerated structure* in the sense of Mal'cev [71] Ch.18. However, for a program of constructive generalization of algebra via formalization in T_0 it should not be necessary to demand of all the structures $(\mathfrak{A},E)$ considered that they be denumerable or decidable. Such additional information is only to be assumed where necessary and verified where possible.

One place where decidability restrictions may play an essential role in algebra is in the ideal theory of rings. For example, a non-trivial ideal X in the integers can only be shown to be principal if it has a least positive element z. But as observed in 4.3.3, this cannot be constructively derived in general. However, if X is decidable with characteristic function g and x is any given positive element in X then we can find z as $(\mu y \leq x)gy \simeq \underline{0}$.[8]

6. T_1 *and related theories*

6.1. Language and axioms. T_1 uses the same language $\mathcal{L}$ as T_0 except that (for simplicity) we adjoin one new constant symbol $\underline{e}_N$. There is only one new axiom:

VI *(Numerical quantification)*

$$(f : N \to N) \to (\underline{e}_N f \simeq \underline{0} \vee \underline{e}_N f \simeq \underline{1}) \wedge (\underline{e}_N f \simeq \underline{0} \leftrightarrow \exists x \eta N.\ fx \simeq \underline{0})$$

6.2. Some consequences. Using the partial minimum operator μ of §3.3(6) we define the *unbounded minimum operator* μ_0 in T_1 by

$$\mu_0 f \simeq \begin{cases} \underline{0} & \text{if } \underline{e}_N f \simeq \underline{1} \\ \mu f & \text{if } \underline{e}_N f \simeq \underline{0}\ . \end{cases}$$

Thus $\mu_0 f\downarrow$ for all $f : N \to N$.

The recursion-theoretic jump operator $J : N^N \to N^N$ is defined as $J(f) \simeq \lambda x.\underline{e}_N(\lambda y.\ t(f,x,y))$ for a certain primitive recursive t. Then the definition of the *hyperarithmetic hierarchy* $\langle H_a \rangle_{a \eta 0_1}$ can be given in T_1, iterating J along 0_1. From this, one defines the predicate $Hyp(f)$ expressing that an operation f in N^N is hyperarithmetic, i.e. recursive in some H_a.

There are several ways to introduce the notion of a function *partial recursive* in $\underline{e}_N$, and to give an associated enumeration

---

[8] This suggests a response to Bishop's question in Bishop [70], pp. 55-56.

$\{z\}^{\underline{e}_N}$ $(z=0,1,2,\ldots)$. One way is by means of Kleene's notion of *partial recursive functional of finite type* in Kleene [59]. This leads to an inductive definition of $\{z\}^{\underline{e}_N}(x_1,\ldots,x_n) \simeq y$ (for $x_1,\ldots,x_n,y$ in N) which falls under Axiom V and can be carried out conveniently in T_0. Kleene shows that

$$\forall f \eta N^N [\mathrm{Hyp}(f) \leftrightarrow \exists z \eta N \; \forall x \eta N (fx \simeq \{z\}^{\underline{e}_N}(x))].$$

Thus the following statement is analogous to Church's Thesis:

(HT) $$\forall f \eta N^N \; \mathrm{Hyp}(f).$$

6.3. Metamathematical results

THEOREM 6.3.1

There is a model of T_1 *in which the range of the variables is the set* ω *of natural numbers and* (HT) *is true.*

By 4.1 Remark (a), this may be proved in exactly the same way as 4.1.1. Note that as in Remark (c) there, this proof can also be formalized in classical 2nd order analysis.

There is an obvious generalization of HEO to any enumerative system. In particular, the enumeration indicated in 6.2 of the functions partial recursive in $\underline{e}_N$, induces a type structure

$$(\mathrm{HEO}_\sigma^{(\underline{e}_N)}, \equiv_\sigma^{(\underline{e}_N)})$$

or as we shall write it $(\mathrm{HEO},\equiv)^{(\underline{e}_N)}$.

COROLLARY 6.3.2

If θ *is a sentence of* $\mathcal{L}_{FT}$ *and* $T_1 \vdash \theta^{(\bar{N},E)}$ *with classical logic then* $\theta^{(HEO,\equiv)^{(e_N)}}$ *is true.*

Without further work we also obtain the results of 4.2.

THEOREM 6.3.3

Theorem 4.2.1 remains correct when "T_0" *is replaced by* "T_1".

COROLLARY 6.3.4

If θ *is a sentence of* $\mathcal{L}_{FT}$ *and* $T_1 \vdash \theta^{(\bar{N},E)}$ *then* $ZF \vdash \theta^{(M)}$.

Turning now to 4.3, define $T_1^{(-)}$ to be $T_1^{(-)}$ + (Axiom VI for e_N). The classes of formulas $\mathcal{F}_0$, $\mathcal{F}_1$ are defined in the same way as before.

THEOREM 6.3.5

Theorem 4.3.1 remains correct when "$T_0^{(-)}$" *is replaced by* "$T_1^{(-)}$".

The proof is as before.

6.4. Relations with predicativity. The informal conception of predicativity taken here is that one deals just with the *definitions and proofs implicit in assuming that the set of natural numbers is given* (as a kind of "completed infinite totality"). Precise proposals for explaining this have been given in terms of *autonomous progressions of ramified theories* R_α ; cf. Feferman [68] for a survey of work on these[9]. Viewed from the outside, the least non-autonomous ordinal is a certain (recursive) Γ_0. The *general concept of ordinal* (or *well-ordering*) is itself *not* predicative. One may speak of *particular ordinals* being predicative when it has been recognized by these means that corresponding principles of transfinite induction and recursion are justified.

The language $\mathcal{L}$ does not match directly with formal languages considered up to now in the study of predicativity. Nevertheless, it makes sense to interpret V as ω and the operations and classifications as ranging over predicative definitions of partial operations and subsets of ω, resp. Clearly, not all of Axiom V for inductive generation is justified under this interpretation. Let V_N be the special case used in 3.3(6) to derive N, and let $T_1^{(N)}$ be T_1 with V_N in place of V.

CONJECTURE

$T_1^{(N)}$ *is (proof-theoretically) reducible to predicative analysis* $\bigcup_{\alpha<\Gamma_0} R_\alpha$.

It may even be that $T_1^{(N)}$ is of the same strength as predicative analysis. (The latter is known to be weaker than the intuitionistic 1st order theory of 0_1, hence also weaker than T_0.)

The actual development of analysis by predicative means may be referred to the hierarchies N_σ or $(\overline{N}_\sigma, E_\sigma)$ in $T_1^{(N)}$. All of classical analysis and much of the modern theory of measure and

[9] Cf. also Feferman [64], Kreisel [70]. A more perspicuous formalization without progressions is given by Feferman in a paper to appear in the Lorenzen Festschrift.

integration can be accounted for predicatively, though the l.u.b. principle is available only for sequences rather than for sets in general[10].

6.5. Borelian mathematics and T_1. Obviously it is necessary to use Axiom V to deal with the parts of mathematics where ordinals enter unrestrictedly. In analysis this shows up in the sequence of derived sets of a closed set, in the theory of Borel sets, etc. Borel and his school (Baire, Lebesgue, etc.) talked of restricting mathematics to that which was explicitly definable (Borel [14]). However, they never made clear what means of definition or proof were to be admitted. Some idea of this can be drawn from their practice, which is seen to be accounted for in T_1. It would be of interest to see whether the *Borelian conception of mathematics* can be explained in a precise way; T_1 would seem to be a strong candidate for this.

The axiom VI with intuitionistic logic implies what Bishop calls the *limited principle of omniscience*; for $f : N \to N$,

(LPO) $$\forall x(fx \simeq 0) \vee \exists x,y(fx \simeq y + 1).$$

He says (Bishop [67], p.9) that each of his results ϕ is a constructive substitute for a classical theorem ψ such that ϕ together with (LPO) implies ψ. Relying on this and 5.1(2) we have:

CLAIM

All of the classical mathematics replaced by Bishop's work can be formalized in T_1.

---

[10] cf. Lorenzen [65] for classical analysis. I have given a predicative development of measure theory in unpublished notes. Transcendental methods in algebraic number theory have also been treated predicatively (Larson [69]).

In most cases this is simpler to verify directly than to pass through this work.

6.6. Relations with hyperarithmetic mathematics. The idea here is the same as for recursive mathematics, and the discussion of §5.2 can be paralleled completely, but the subject itself has not been pursued to anywhere near the same extent. For the most part the positive results have already been realized as predicative theorems, hence as generalizations of both hyperarithmetic and classical results. By 6.3.2, negative results such as that of Kreisel [59] for the Cantor-Bendixson theorem may be used to give independence of some classical theorems from T_1.

There is one recent positive development in hyperarithmetic mathematics that ought to be re-examined with an eye to generalization by means of formalization in T_1, namely *hyperarithmetic model theory*. This was initiated by Cleave [68] particularly in a study of hyperarithmetic analogues of ultrafilters, ultrapowers, etc. Cutland [72] has continued this for saturated models and forms of categoricity. Denumerable models should play a special role since for these the satisfaction relation is decidable.

7. *Concluding questions and remarks*

7.1. Systematic and ad hoc explicit mathematics. What we have called here *systematic* explicit mathematics are attempts to redevelop substantial portions of mathematics by means of restricted methods of definition and/or proof. By contrast, *ad hoc* work examines particular existential results of classical mathematics with the aim to obtain more explicit or sharper information. No (deliberate) restriction is made on methods of definition or proof; rather one employs refined considerations or special new methods. The most striking (and frequently cited)[11] example is Baker's work on some classes of diophantine equations, giving explicit bounds for the solutions which had previously only been known to be finite in number. Taking a systematic approach will not automatically lead one to such improvements.

NOTE

Explicit definability results for theories with logic restricted to be intuitionistic give them an appearance of systematic explicit mathematics. Here the emphasis instead has been on the choice of basic notions to more accurately reflect actual practice. Intuitionist logic plays a role only in some metatheorems.

7.2. Proof-theoretical work on subsystems.

For certain subtheories T of T_0, T_1 it has been possible to characterize the operations from N to N which are proved in T to exist. These characterizations are given in terms of certain hierarchies of functions up to some familiar ordinals. The techniques are proof-theoretical, by Gödel's functional interpretation (Gödel [58]) followed by normalization of terms. This has been done for systems of finite type over N in Tait [65] and, with $\underline{e}_N$, in Feferman [70]. Howard [72] treats a system of finite type over N, 0_1 which is interpretable in T_0, and Zucker (Troelstra [72]) gives some extensions of this to iterated inductive definitions; it should

---

[11] Particularly by Kreisel; cf. Kreisel [74] for a more adequate discussion.

be possible to use similar methods when the axiom VI for $\underline{e}_N$ is added. It may be of interest to see whether these techniques can be applied directly to related and stronger subsystems of T_0 and T_1 with "variable types". Some such has been done by Girard [71] and Martin-Löf [Ms] .

7.3. *Bounded classifications and sets*

The following kind of extension S_0 of T_0 might provide a more flexible comparison with set theory. The language $\mathcal{L}_{Bd}$ of S_0 has one new basic predicate symbol Bd(x) which we read as: x *is a bounded classification*; the idea is that x is contained in some classification built up by sums and products from N. The axioms of S_0 agree with those of T_0 through I-IV; V is modified so as to allow any formula ψ of $\mathcal{L}_{Bd}$. We have in addition the following axioms for Bd:

(i) $Bd(x) \to Cl(x)$

(ii) $Bd(N)$

(iii) $Bd(A) \wedge \forall x \eta A.\ Bd(fx) \to Bd(\Sigma_{x\eta A} fx) \wedge Bd(\Pi_{x\eta A} fx)$

(iv) $Bd(A) \wedge X \subseteq A \to Bd(X)$

(v) A scheme for proof by induction on Bd.

Since $A \times B \subseteq \Sigma_{x\eta A} B$ and $B^A \subseteq \Pi_{x\eta A} B$, we also have closure of Bd under these operations. Note that if $\sigma\eta Typ_N$ then $Bd(N_\sigma)$, $Bd(\overline{N}_\sigma)$ and $Bd(E_\sigma)$; more generally these hold for $\sigma\eta Typ_A$ where each X in A is bounded.

S_0 + (CT) has a model just like that for T_0 in 4.1.1. Simply take Bd to be the smallest set satisfying (ii)-(iv). The same method works in any model, where i(a,r) is set-theoretically defined by:

$$z \eta i(a,r) \Leftrightarrow z \in \cap X\{\forall x \eta a[\forall y(y,x)\ r \Rightarrow y \in X) \Rightarrow x \in X]\}$$

QUESTION

Is S_0 *a conservative extension of* T_0 *?*

We can associate with each model of set theory a model of S_0 in which the bounded classifications are just those coextensive with the sets:

THEOREM 7.3.1

Let $\mathfrak{U} = (A,\in)$ *be any model of* ZF. *There is a model* $\mathfrak{U}^*$ *of* S_0 *in which* $V = A$ *and* $Bd(a) \Leftrightarrow Cl(a) \wedge \exists b \in A\ \forall x \in A[x \eta a \Leftrightarrow x \in b]$.

The method of proof is the same as for 4.2.1.

For a nice result like 4.2.2 one should perhaps deal instead with axioms for the predicate Set(A,E) which holds when Bd(A) and E is an equivalence relation on A.

NOTE

Bishop's notion of set in Bishop [67] may be more appropriately interpreted by such a predicate.

7.4. *Impredicatively defined operations*

In analogy with the introduction of $\underline{e}_N$ we might further consider the theory T_2 obtained from T_1 by adding a constant $\underline{e}_{N^N}$ with the axiom:

$$(f : N^N \to N) \to (e_{N^N}f \simeq \underline{0} \vee e_{N^N}f \simeq \underline{1}) \wedge [e_{N^N}f \simeq 0 \leftrightarrow \exists g \eta N^N . fg \simeq \underline{0}].$$

Mathematically, this permits application of the l.u.b. principle to decidable sets of real numbers. Questions here would be whether we could get a model like that in 6.3.1 for T_1, and to what extent we could get interesting proof-theoretical information about subsystems of T_2 including the new axiom.

7.5. *Impredicatively defined classifications*

Here we would like to know to what extent T_0 can be strengthened by classification existence axioms so that the resulting theory T^* is also intuitively correct, or at least for which we can get a model of $T^* + (CT)$ in which N is standard. Particularly to be considered are instances of the *comprehension axiom scheme*

$$(CA_\phi) \qquad \exists C\ \forall x[x\eta C \leftrightarrow \phi(x)]$$

where ϕ is not an elementary formula. The fact that the unrestricted application of this principle leads to contradiction shows that the concept of classification is not completely clear; cf. §7.7 below. Some experimentation to see how far the use of (CA) can be pushed may be helpful to obtain clarification in these circumstances. The specific cases considered in this section and the next were suggested by past experience.

Call ϕ *2nd order* if it is a formula of $\mathcal{L}^{(2)}$ satisfying the conditions to be an elementary formula (2.6) except that quantifiers with classifications variables $\forall X(\ldots)$, $\exists X(\ldots)$ are permitted. Write $\phi^{(\subseteq A)}$ for the result of replacing each such quantifier by $\forall X \subseteq A(\ldots)$, $\exists X \subseteq A(\ldots)$.

It turns out possible to get a model of $T_0 + (CT) +$

$(CA_\phi(\underline{\subset}N))_\phi$ 2nd order in which N is standard. This can be done more generally with N replaced by any bounded classification. Let $(CA^{(2)})$ be the schema (CA_ϕ) for all 2nd order ϕ.

QUESTION

Does $T_0 + (CT) + (CA^{(2)})$ *have a model, particularly one in which* N *is standard?*[12]

Note that the inductive generation axiom V is derivable from $(CA^{(2)})$.

7.6. Impredicatively defined classifications (cont.)

Another collection of instances of the scheme (CA) which should be considered is suggested by *self-applicable concepts in algebra* such as the "category of all categories". In T_0, (single-sorted) structures of signature $\nu = ((n_1,\ldots,n_k), (m_1,\ldots,m_\ell), (m_1,\ldots,m_\ell),p)$ are defined to be $(1 + k + \ell + p)$-tuples:

(i) $a = (A,R_1,\ldots,R_k, f_1,\ldots,f_\ell, c_1,\ldots,c_p)$

where

(ii) each $R_i \subseteq A^{n_i}$, $f_i : A^{m_i} \to A$, and $c_i \eta A$.

Write $Str_\nu(a)$ for the formula in $\mathcal{L}$ which expresses that there exist $A,R_1,\ldots,c_p$ satisfying (i), (ii). Given any sentence θ in the 1st order language $\mathcal{L}_\nu^{(1)}$ of structures of type ν, write $Sat_\theta(a)$ for the formula which expresses that the structure a *satisfies* (or is a *model* of) θ ; we also write $a \models \theta$ for this. Finally, write $a_1 \cong a_2$ to express that a_1, a_2 are isomorphic structures of the same signature.

The first question would be whether we can consistently assume for each ν : $\exists B\, \forall x[x \eta B \leftrightarrow Str_\nu(x)]$. This is not possible

[12] (Added in proof) The answer is positive; cf. Addenda (A2) below.

with T_0 for then we could derive the existence of a C such that $\forall x[x\eta C \leftrightarrow C\ell(x)]$. A form of Russell's paradox follows by taking $J = \Sigma_{x\eta C} x$ and $D = \hat{x}.(x,x)\not\eta J$. But this does not exclude the possibility of having classes of *representatives* (with respect to $\cong$) of all structures of a given type, which is all that is important algebraically.

QUESTION

Is T_0 *consistent with the following sentences*

$$\exists C[\forall x(x\eta C \rightarrow Str_\nu(x) \wedge x \models \theta) \wedge \forall y(Str_\nu(y) \wedge y \models \theta \rightarrow \exists x\eta C(y \cong x))]$$

for each signature ν *and sentence* θ *of* $\mathcal{L}_\nu^{(1)}$*?*

Perhaps more promising is the use of $T_0 + (CA^{(2)})$ of the preceding section. We can already speak in that theory (to be more precise, in a slight extension) of the category of all functors between any given "large" categories (e.g. groups, classes, etc.): for, functors are just operations satisfying special conditions explained by quantifying over structures.

REMARK

The systems developed in (Feferman [Ms]) could deal with all these kinds of self-applicable concepts, but at a cost of other deficiencies. A principal difference is that operations were explained there in terms of classes rather than treated independently as here. The present language should permit greater flexibility.

7.7. *Partial and total classifications.*

Returning to the question of clarifying the concept Cl, one

way of putting the difficulty is that not every well-formed formula $\phi(x,\ldots)$ of $\mathcal{L}$ need be recognized as determining a property which is *meaningful for all* x. For example, it might be said that the property of being a Cauchy sequence is only meaningful for sequences. This suggests considering a notion of *partial property* or *classification*, one whose *domain of significance* may be only a part A of V. We would only be able to say in this case:

$$\exists C\ \forall x\eta A\ [x\eta C \leftrightarrow \phi(x)].$$

This appealing idea goes back to Russell; one form of it has been pursued by Gilmore [70]. Another point that might appeal in the present context is that it appears to put operations and classifications on a similar footing.

If T_0 is to be embedded in a theory of partial classifications it seems we should have a new operation δ such that

$$Cl(x) \rightarrow \delta x\downarrow \wedge Cl(\delta x).$$

Here $Cl(x)$ is read as: x is a partial classification, and δx is read as: the domain of significance of x. We would call x a *total* classification if $V \subseteq \delta x$; it is such that we have had in mind up to this section. In a theory of this kind, the partial comprehension scheme would take the form for arbitrary ϕ :

$$\exists C\ \forall x\eta\delta C\ [x\eta C \leftrightarrow \phi(x)].$$

If nothing more is said about the members of C, this theory is trivially consistent. The problem with this idea is that we have shifted the initial question to: which ϕ have $V \subseteq \delta C$? One would hope to get simple evident (sufficient) conditions for this, to recapture at least axioms III-V, if not more[13].

7.8. *Perspective*

The study of systematic explicit mathematics may be more of logical and/or philosophical interest than of mathematical interest. In any case it is relevant to significant portions of actual mathematics. The problems raised in 7.5 - 7.7 are very intriguing from the logical point of view, but they have little mathematical relevance, as far as one can see now.

ADDENDA

(A1) To Footnote 2.

Following circulation of this paper I learned of a theory CST of functions and sets independently developed by Myhill, which has several aspects in common with T_0. His stated purpose is to provide a constructive framework for constructive mathematics as exemplified in Bishop [67]. CST differs from T_0 in the basic respect that extensionality for both functions and sets is taken among the axioms. For this reason, CST is not evidently constructive. It is possible though that Myhill's metamathematical work on the theory will show it reducible to constructive principles. (This is being prepared for publication.)

I also learned from Myhill of the paper by Cocchiarella [to appear] which introduces some axioms for predicates and corresponding sets (here: classifications) that may be said to anticipate the idea for

[13](Added in proof) Subsequent to the above I found some theories of partial operations and classifications which accomplish a good deal of this aim as well as that of 7.6. The work will appear in a paper for the Schütte Festschrift.

axiom schema III of T_0 (3.3 below).

(A2) To §7.5.

This gives an affirmative answer to the question raised in §7.5 whether there is a model $\mathfrak{N}$ for T_0 + (CT) with full 2nd order comprehension $CA^{(2)}$ (and in which N is standard). The proof is by a non-constructive modification of the proof of 4.1.1.
Let $\mathfrak{M} = (\omega, \mathcal{P}\omega, \simeq, \in)$; the interpretations of $0, k, s, d, p, p_1, p_2$ are chosen as before. Associate with each 2nd order formula $\phi(x_1,\ldots,x_n, X_1,\ldots,X_m, Y)$ (where the variables listed include all free variables of ϕ) a Skolem function $Y = F_\phi(x_1,\ldots,x_n,X_1,\ldots,X_m)$ in $\mathfrak{M}$. Now define the subsets $C\ell_\alpha$ of ω and for $a \in C\ell_\alpha$ the set $e(a) = \{x : x\eta a\}$ by induction on α ($e(a)$ is the *extension* of a) as follows. $C\ell_0 = 0$ and for limit λ, $C\ell_\lambda = \cup_{\alpha<\lambda} C\ell_\alpha$. $C\ell_{\alpha+1}$ is $C\ell_\alpha$ together with

(i) all $(1,\phi,(x_1,\ldots,x_n,a_1,\ldots,a_m))$ where ϕ is 2nd order as above and each $a_1,\ldots,a_m \in C\ell_\alpha$, as well as

(ii) all $(2,a,f)$ such that $a \in C\ell_\alpha$ and $\forall x\eta a[\{f\}(x) \in C\ell_\alpha]$.

In case (i), take $z\eta(1,\phi,(x_1,\ldots,x_n,a_1,\ldots,a_m)) \Leftrightarrow z \in F_\phi(x_1,\ldots,x_n, e(a_1),\ldots,e(a_m))$ and in (ii), take $z\eta(2,a,f) \Leftrightarrow \exists x\eta a\, \exists y\eta\{f\}(x) [z = (x,y)]$. Let $C\ell = \cup_\alpha C\ell_\alpha$ and $N = (\omega, C\ell, \simeq, \eta)$.

LEMMA

For any 2nd order $\psi(x_1,\ldots,x_n,X_1,\ldots,X_m)$ *(considered as a formula of* $\mathcal{L}$*) and any* $x_1,\ldots,x_n \in \omega$, $a_1,\ldots,a_m \in C\ell$ *we have*
$\mathfrak{N} \models \psi(x_1,\ldots,x_n,a_1,\ldots,a_m) \Leftrightarrow \mathfrak{M} \models \psi(x_1,\ldots,x_n,e(a_1),\ldots,e(a_m))$.

Thus $\mathfrak{N}$ is a model of $(CA^{(2)})$ + (CT) as well as axioms I, II, IV of T_0. The verification of $(CA^{(2)})$ takes care of III

(Elementary Comprehension) and V (Inductive generation).

In the same way we can strengthen Theorem 6.3.1.

* * * * * *

BIBLIOGRAPHY

BARWISE, J.

[69] Infinitary logic and admissible sets, J.Symb. Logic 34, 1969, 226-252.

BISHOP, E.

[67] Foundations of Constructive Analysis, McGraw-Hill, New York, 1967.

[70] Mathematics as a Numerical Language, in Intuitionism and proof theory (eds. Kino, Myhill, Vesley), North-Holland, Amsterdam, 1970.

BISHOP, E. and H. CHANG

[72] Constructive measure theory, Memoirs A.M.S. no. 116, 1972.

BOREL, E. (et al)

[14] Cinq lettres sur la théorie des ensembles, repr. in Leçons sur la théorie des fonctions, 2nd ed., Gauthier-Villars, Paris, 1914, 150-160.

CLEAVE, J.P.

[68] The theory of hyperarithmetic and Π^1_1-models, in Proc. Summer School in Logic, Lecture Notes in Maths. 70, Springer, Berlin, 1968, 223-240.

COCCHIARELLA, N.B.

[to appear] Formal ontology and the foundations of mathematics, in the Philosophy of Bertrand Russell: A Centenary Tribute (ed. Nakhnikian), to appear.

CROSSLEY, J.N.

[69] Constructive Order Types, North-Holland, Amsterdam, 1969.

CUTLAND, N.

[72] Π_1^1-models and Π_1^1-categoricity, in Conf. in Math. Logic, London '70, Lecture Notes in Maths. 255, Springer, Berlin, 1972, 42-61.

DEKKER, J.C.E. and J. MYHILL

[60] Recursive Equivalence Types, U. Calif. Publs. in Maths. 3, 1960, 67-214.

ERSHOV, Yu.L.

[68] Numbered fields, in Logic, Methodology and Philosophy of Science III (eds. van Rootselaar, Staal), North-Holland, Amsterdam, 1968.

FEFERMAN, S.

[64] Systems of predicative analysis, J. Symb. Logic 29, 1964, 1-30.

[68] Autonomous transfinite progressions and the extent of predicative mathematics, in Logic, Methodology and Philosophy of Science III, North-Holland, Amsterdam, 1968.

[70] Ordinals and functionals in proof theory, in Proc. Int. Cong. Maths., Nice, 1970, V.1, 229-233.

[Ms] Some formal systems for the unlimited theory of structures and categories, unpublished Ms.

FRIEDMAN, H.

[71] Axiomatic recursive function theory, in Logic Colloquium '69 (eds. Gandy, Yates), North-Holland, Amsterdam, 1971, 113-137.

GILMORE, P.C.

[70] Attributes, sets, partial sets and identity, Compos. Math. 20, 1960, 53-79.

GIRARD, J.-Y.

[71] Une extension de l'interpretation de Gödel à l'analyse, et son application a l'élimination des coupures dans l'analyse et la théorie des types, in Proc. 2nd Scand. Logic Sympos. (ed. Fendstadt), North-Holland, Amsterdam, 1971, 63-92.

GÖDEL, K.

[58] Über eine bisher noch nicht benutzte Erweiterung des finiten Standpunktes, Dialectica 12, 1958, 280-287.

GOODMAN, N. and J. MYHILL

[72] The formalization of Bishop's constructive mathematics, in Toposes, algebraic geometry and logic, Lecture Notes in Maths. 274, Springer, Berlin, 1972, 83-96.

HEYTING, A.

[72] Intuitionism, an introduction, (3rd rev. edn.), North-Holland, Amsterdam, 1972.

HOWARD, W.A.

[72] A system of abstract constructive ordinals, J. Symb. Logic 37, 1972, 355-374.

KLEENE, S.C.

[52] Introduction to metamathematics, Van Nostrand, New York, 1952.

[59] Recursive functionals and quantifiers of finite types I, Trans. A.M.S. 91, 1959, 1-52.

KREISEL, G.

[59] Analysis of the Cantor-Bendixson theorem by means of the analytic hierarchy, Bull. Pol Acad. Sci. 7, 1959, 621-626.

[65] Mathematical logic, in Lectures in Modern Mathematics,III, (ed. Saaty), Wiley, New York, 1965, 95-195.

[70] Principles of proof and ordinals implicit in given concepts, in Intuitionism and proof theory, North-Holland, Amsterdam, 1970.

[71] Some reasons for generalizing recursion theory, in Logic Colloquium '69 (eds. Gandy, Yates), North-Holland, Amsterdam, 1971, 139-198.

KREISEL, G.

[to appear] From proof theory to a theory of proofs, 1974 AAAS Symp., to appear.

KREISEL, G. and A.S. TROELSTRA

[70] Formal systems for some branches of intuitionistic analysis, Annals Math. Logic 1, 1970, 229-387.

LACOMBE, D.

[59] Quelques procédés de définition en topologie récursive, in Constructivity in Mathematics (ed. Heyting), North-Holland, Amsterdam, 1959, 129-158.

LARSON, A.

[69] A case study of predicative mathematics: uniform spaces, completions and algebraic number theory. Dissertation, Stanford University, 1969.

LORENZEN, P.

[65] Differential and Integral: Eine konstruktive Einführing in die klassische analysis, Akad. Verlag, Frankfurt,1965.

MAL'CEV, A.I.

[71] The metamathematics of algebraic systems; collected papers: 1936-1967, (transl., ed. Wells), North-Holland, Amsterdam, 1971, Chps. 18, 22, 24.

MARTIN-LÖF, P.

[preliminary Ms] An intuitionistic theory of types.

MOSCHOVAKIS, Y.N.

[69] Abstract first order computability, I, Trans. A.M.S. 138, 1969, 427-464.

RABIN, M.

[62] Computable algebra; general theory and theory of commutative fields, Trans. A.M.S. 95, 1962, 341-360.

SANIN, N.A.

[68] Constructive real numbers and function spaces, 21, Translations of Mathematical Monographs, A.M.S., Providence, 1968.

SCOTT, D.

[70] Constructive validity, in Symposium on Automatic Demonstration, Lecture Notes in Maths. 125, Springer, Berlin, 1970, 237-275.

SPECKER, E.

[59] Der Satz vom Maximum in der rekursiven Analysis, in Constructivity in Mathematics, (ed. Heyting), North-Holland, Amsterdam, 1959, 254-265.

TAIT, W.W.

[65] Infinitely long terms of transfinite type, in Formal Systems and Recursive Functions, (eds. Crossley, Dummett), North-Holland, Amsterdam, 1965, 176-185.

[68] Constructive reasoning, in Logic, Methodology and Philosophy of Science III (eds. van Rootselaar, Staal), North-Holland, Amsterdam, 1968.

TROELSTRA, A.S.

[69] Principles of intuitionism, Lecture Notes in Maths. 95, Springer, Berlin, 1969.

[73] Metamathematical investigation of intuitionistic arithmetic and analysis, Lecture Notes in Maths. 344, Springer, Berlin, 1973.

TSEYTIN, G.D., I.D. ZASLAVSKY and N.A. SHANIN

[66] Peculiarities of constructive mathematical analysis, in Proc. Int. Cong. Maths., Moscow, 1966, 253-261.

ZUCKER, J.I.

[73] Iterated inductive definitions, trees and ordinals, in (Troelstra [73]), Ch. VI, 392-453.

* * *

Department of Mathematics, Stanford University, Stanford, California.

DIMENSION THEORY OF COMMUTATIVE POLYNOMIAL RINGS

Robert Gilmer

In this talk I shall present a survey of some known results concerning the Krull dimension of polynomial rings in finitely many indeterminates over a commutative ring with identity.

Throughout the talk the letter R designates a commutative ring with identity. A basic concept considered is the (Krull) *dimension* of a ring, defined as follows: if the nonnegative integer n is such that there exists a chain

$$P_0 \subset P_1 \subset P_2 \subset \dots \subset P_n$$

of proper prime ideals of R of length n, but no such longer chain, then R has dimension n; if no such nonnegative integer n exists (that is, if there exist arbitrarily large finite chains of proper prime ideals of R), then the dimension of R is infinite[1]. The questions considered in the talk are of interest only in the case of finite-dimensional rings, and hence I assume throughout that R has finite dimension n_0; the symbol dim S stands for the dimension of the ring S.

---

[1]Sometimes one distinguishes between different infinite cardinals in defining the dimension of a ring (for example, see Bass [1971]), but there is no reason to do so here. For rings without identity with no proper prime ideals (that is, rings in which each element is nilpotent), the dimension is defined to be -1.

For a positive integer k, the symbol $R^{(k)}$ denotes the polynomial ring $R[X_1,\ldots,X_k]$ in k indeterminates over R. In Seidenberg [1953], A. Seidenberg established the inequalities

$$n_0 + 1 \leq \dim R^{(1)} \leq 2n_0 + 1.$$

Consequently, the ring $R^{(k)}$ is finite-dimensional for each posite integer k; if $n_k = \dim R^{(k)}$, then the sequence $\{n_i\}_{i=0}^{\infty}$ is called the *dimension sequence* of R and the sequence $\{n_i - n_{i-1}\}_{i=1}^{\infty}$ is called the *difference sequence* of R (Bastida and Gilmer [1973]). This talk deals with the problem of determining those sequences $\{a_i\}_{i=0}^{\infty}$ of nonnegative integers that are dimension sequences, and I shall frequently refer to this as 'the dimension sequence problem'. I should state in the beginning that a final resolution of this question has been made within the last two years; the work I have in mind is by J.T. Arnold and myself, and will appear as Arnold and Gilmer [1973], but the process of solving the problem has taken, as you will see, more than twenty years. There are three aspects to the problem of determining all dimension sequences; I label these (I), (II), and (III).

(I) Determine conditions that a dimension sequence satisfies.

(II) Formulate a conjecture as to what sequences are dimension sequences.

(III) The realization problem; that is, the problem of showing that a sequence can be realized as a dimension sequence.

As might be expected, initial contributions to the dimension sequence problem concerned (I) – conditions that a dimension sequence satisfies. In his historic 1951 paper (Krull [1951]) on Hilbert's Nullstellensatz and what have come to be commonly called *Hilbert rings* (*Jacobsonsche Ringe* in Krull [1951]), W. Krull proved that $\dim R^{(k)} = \dim R + k$ for all k, if the ring R is Noetherian. Actually, Krull was not directly concerned with the dimension sequence problem in Krull [1951], and you must look beyond statements of results to find the theorem I have just stated. At this stage I should mention the *dimension sequence problem for rings with property* E — that is, the problem of determining the set of dimension sequences of rings with property E. Krull's result resolves, of course, the dimension sequence problem for Noetherian rings; they are the sequences $\{m, m+1, m+2, \ldots\}$, where m is any nonnegative integer.

In 1953 and 1954, Seidenberg published two papers (Seidenberg [1953, 1954]) dealing with the dimension theory of rings. As stated earlier, he established the inequalities.

$$\text{(R1)} \qquad n_0 + 1 \leq \dim R^{(1)} \leq 2n_0 + 1$$

In particular, a dimension sequence is strictly increasing — that is, the difference sequence of a ring is a sequence of positive integers. Seidenberg also proved that the bounds given in (R1) are the best possible in the sense that for each pair (n_0, n_1) of nonnegative integers such that $n_0 + 1 \leq n_1 \leq 2n_0 + 1$, there exists an integral domain D^* such that $\dim D^* = n_0$ and $\dim D^*[X] = n_1$. Result (R1) implies that if R is one-dimensional, then $\dim R^{(1)}$ is 2 or 3. For a one-dimensional integral domain R, Seidenberg proved that $\dim R^{(1)} = 2$ if and only if the integral closure of R is a

Prüfer domain[2]; moreover, if J is a Prüfer domain, then $\dim J^{(k)} = \dim J + k$ for each k, and hence the set of dimension sequences for Prüfer domains is the same as for Noetherian rings — they are the sequences of nonnegative integers with corresponding difference sequence $\{1,1,1,...\}$. Note that Seidenberg's results imply that the sequence $\{1,2,3,...\}$ is the only sequence with first two members 1, 2 that can be realized as the dimension sequence of an integral domain.

Seidenberg's work was followed in 1960 by the monograph (Jaffard [1960]) of P. Jaffard that, as its title states, is devoted to an investigation of the dimension theory of polynomial rings. Using the fact that $\dim K^{(n)} = n$ for all n if K is a field, Jaffard observed that for a fixed proper prime ideal P of R, a chain of prime ideals of $R^{(m)}$, each lying over P in R, contains at most $m+1$ members. In turn, this observation implies

$$\text{(R2)} \qquad n_0 + k \leq \dim R^{(k)} \leq (n_0+1)(k+1)-1 \quad \text{for all } k.$$

As with Seidenberg's result (the special case of (R2) with $k = 1$), the bounds given in (R2), for a fixed k, cannot be sharpened. If $n_0 = 0$, then the upper and lower bounds in (R2) are equal to k, and hence each zero-dimensional ring has dimension sequence

---

[2] An integral domain D with identity is a *Prüfer domain* if each nonzero finitely generated ideal of D is invertible; the term honours H. Prüfer, who considered such domains in Prüfer [1932], under the designation 'a domain with property $\mathcal{L}$B'. Numerous characterizations of such domains, within the class of integral domains with identity, are known (see, for example, Chapter IV of Gilmer [1972]); three such characterizations are
(1) an integral domain D with identity is a Prüfer domain if and only if D_P is a valuation ring for each proper prime ideal P of D,
(2) D is a Prüfer domain if and only if the set of ideals of D forms a distributive lattice under $+$ and $\cap$, and
(3) D is a Prüfer domain if and only if each overring of D is integrally closed.

$\{0,1,2,\ldots\}$. In particular, a (von Neumann) regular ring has dimension sequence $\{i\}_{i=0}^{\infty}$. What are the dimension sequences for one-dimensional rings? Jaffard showed that there are essentially three possibilities, as follows:

dimension sequence	difference sequence
$\{1,2,\ldots,n,\ldots\}$	$\{1,1,\ldots,1,\ldots\}$
$\{1,3,\ldots,2n+1,\ 2n+2,\ldots\}$	$\{2,2,\ldots,2,1,1,\ldots\}$
$\{1,3,\ldots,2n+1,\ldots\}$	$\{2,2,\ldots,2,\ldots\}$

He also showed that each of the preceding sequences can be realized as the dimension sequence of an integral domain. It follows that the first r terms (for any r) of a dimension sequence do not, in general, determine the dimension sequence.

Based upon the results stated to this point and, I must admit, upon a bit of hindsight, the following four questions seem reasonable[3].

(1) Is the difference sequence of a ring non-increasing?

(2) Is the difference sequence of an n_0-dimensional ring bounded above by $n_0 + 1$?

(3) Is the difference sequence of a ring eventually constant?

(4) Is $\mathcal{D}_R$, the set of dimension sequences of rings, the same as $\mathcal{D}_I$, the set of dimension sequences of integral domains?

It is quite easy, however, to see that the answer to question (1) is negative. Thus, if $\{R_i\}_{i=1}^{n}$ is a finite family of

---

[3]At least if we restrict to the classes of rings heretofore mentioned – Noetherian rings, Prüfer domains, and rings of dimension at most 1 – then the answer to each of the four questions is affirmative.

rings (commutative and with identity, of course) and if $R = R_1 \oplus \ldots \oplus R_n$ is the direct sum of this family, then $\dim R = \sup\{\dim R_i\}_{i=1}^n$; this equality depends upon the known prime ideal structure of R. Moreover, $R^{(k)} \simeq R_1^{(k)} \oplus \ldots \oplus R_n^{(k)}$ for each k, and consequently, $\dim R^{(k)} = \sup\{\dim R_i^{(k)}\}_{i=1}^n$. In other words, the dimension sequence for R is the supremum, in the cardinal order, of the dimension sequences for $R_1, R_2, \ldots, R_n$. From this observation we draw two conclusions, one negative and one positive. The positive conclusion is that $\mathcal{D}_R$ is closed under taking suprema, in the cardinal order, of finite sets, and the negative result is that a difference sequence need not be nonincreasing. For example, $s = \{2,3,4,\ldots\}$ (by Krull) and $t = \{1,3,5,\ldots\}$ (à la Jaffard) are dimension sequences so that $u = \sup\{s,t\} = \{2,3,5,\ldots\}$ is also a dimension sequence, but the difference sequence of u is $\{1,2,2,\ldots\}$.

Jaffard proved that the answer to question (3) is affirmative: the difference sequence of a ring is eventually constant. I consider this to be one of two profound results involved in the solution of the dimension sequence problem. To obtain this theorem, Jaffard introduced a new concept, that of *valuative dimension*, and he proved a somewhat technical result that has come to be known as the Special Chain Theorem:

(R3) The dimension of $R^{(k)}$ can be realized as the length of a chain $\mathcal{C}$ of prime ideals such that for each ideal P in the chain, $(P \cap R) \cdot R^{(k)}$ is also in the chain.

While the Special Chain Theorem has not proved to have wide applicability outside the context in which Jaffard considered it, the notion of valuative dimension has been widely used in commutative algebra; the definition is as follows. If D is an integral domain with quotient field K, then the valuative dimension of D (denoted $\dim_v D$) is defined to be

$$\sup \{\dim V \mid V \text{ is a valuation overring of } D\};$$

here *valuation overring* of D means a valuation ring that is a subring of K containing D. If the set $\{\dim V \mid \ldots\}$ is not bounded above by an integer, then we say that D has infinite valuative dimension. Some results of Krull in Krull [1932] imply that $\dim D \leq \dim_v D$ for each domain D; among the results on valuative dimension established by Jaffard in Jaffard [1960] are two that are especially relevant to the dimension sequence problem[4]:

(R4) $\dim D^{(n)} = \dim D + n$ for all n if and only if $\dim D = \dim_v D$.

(R5) The eventual value of the difference sequence for D is 1 if and only if the valuative dimension of D is finite.

We return to question (2): Is the difference sequence for R bounded above by $n_0 + 1$? Jaffard showed that the eventual value of the difference sequence is bounded above by $n_0 + 1$, and Arnold and I proved that the other terms of the difference sequence are bounded above by $n_0 + 1$ as well.

The following theorem is a summary of what I have stated in regard to question (I).

(R6) Each element of $\mathcal{D}_R$ is an increasing sequence; $\mathcal{D}_R$ is closed under taking suprema of finite sets in the cardinal

---

[4]For some other results on valuative dimension, see Sections 30 and 32 of Gilmer [1972].

order. If $\{n_i\}_0^\infty$ is in $\mathcal{D}_R$ and if $\{d_i\}_1^\infty$ is the difference sequence associated with $\{n_i\}_0^\infty$, then $\{d_i\}_i^\infty$ is bounded above by $n_0 + 1$ and is eventually constant.

There seems to have been no further progress towards the solution of the dimension sequence problem, *per se*, until the appearance of the paper (Bastida and Gilmer [1973]) in 1973 , but we will see that results of the 1969 papers (Arnold [1969]) and (Gilmer [1969]) were used in the problem's eventual resolution. The paper (Bastida and Gilmer [1973]) is concerned with the so-called (D+M)-construction; since I have previously extolled the virtues of the construction in print (Gilmer [1973], pp. 95-96) and (Gilmer [1972], pp. 582-584), I will restrain myself here and stick to giving definitions and stating results pertinent to the topic at hand. Assume that V is a valuation ring of the form $K + M$, where K is a field and M is the maximal ideal of V. If D is a subring of K with identity, then $D_1 = D + M$ is an integral domain with identity and V is a valuation overring of D_1; the domain D_1 is called a (D+M)-construction. Corollary 5.5 of Bastida and Gilmer [1973] gives the precise relationship between the dimension sequences for D and D_1:

(R7) Let d be the transcendence degree of K over F, the quotient field of D. Then

$$\dim D_1^{(i)} = \dim D^{(i)} + r + i \quad \text{for} \quad 0 \leq i \leq d, \text{ and}$$

$$\dim D_1^{(i)} = \dim D^{(i)} + r + d \quad \text{for} \quad i > d.$$

The proof of (R7) uses strongly a result of Arnold in Arnold [1969] that is of interest in its own right.

(R8) Let J be a finite-dimensional integral domain with

identity and with quotient field K and let m be a positive integer. Then

$$\dim J^{(m)} = m + \sup \{\dim J[t_1,\ldots,t_m] \mid \{t_i\}_{i=1}^{m} \subseteq K\}.$$

Because of the latitude one has in choosing r and d (d may be infinite), (R7) provides a powerful device for obtaining new dimension sequences from old ones. For example, beginning with the dimension sequence

$s = \{1,2,3,\ldots\}$, we obtain

$s_1 = \{2,4,6,8,9,10,\ldots\}$ as a dimension sequence, using $r = 1$ and $d = 3$. (I am implicitly using the fact that if L is a field and k is a nonnegative integer, then there is a valuation ring of rank k of the form $L + M$.)

Then from s_1 we obtain

$s_2 = \{3,6,9,11,12,13,\ldots\}$ as a dimension sequence ($r = 1$, $d = 2$), etc. More generally, (R7) and an inductive proof yield the following result concerning the realization problem (III).

(R9) Let $\mathcal{S}$ be the set of strictly increasing sequences $\{n_i\}_0^\infty$ of nonnegative integers such that the associated difference sequence $\{d_i\}_1^\infty$ is bounded above by $n_0 + 1$, non-increasing, and eventually constant. Then $\mathcal{S}$ is contained in $\mathcal{D}_I$, the set of dimension sequences of integral domains.

On the other hand, it is easy to see that if D and D_1 are as in (R7), then the difference sequence for D_1 is non-increasing if the difference sequence for D is non-increasing. Hence (R9) gives rise to the following additional questions.

(5) Is $\mathcal{S}$ equal to $\mathcal{D}_I$?

(6) If $\mathcal{D}$ is the set of sequences obtained as suprema, in the cardinal order, of finite subsets of $\mathcal{S}$, then $\mathcal{D}$ is contained in $\mathcal{D}_R$. Are these two sets equal?

Note that questions (5) and (1) are the same, except for the restriction in (5) to integral domains. Although (5) would seem to be a reasonable question based upon what I have said up to this point, at the time (R9) was proved in 1972, the answer to (5) was known to be negative. In fact, the paper Arnold [1969] that contains (R8) also contains a class of examples, due to W. Heinzer, that show that for each integer n greater than 1, the sequence $\{n, n+1, \ldots, 2n-1, 2n+1, 2n+2, \ldots\}$, with difference sequence $\{1, 1, \ldots, 1, 2, 1, \ldots\}$ is the dimension sequence of an integral domain.

The answer to question (6) is affirmative – that is, $\mathcal{D}_R = \mathcal{D}$. After a couple of false starts, I think that the conjecture that $\mathcal{D}_R$ is equal to $\mathcal{D}$ was formulated late in 1971, with the proof following several months later. The theorem that $\mathcal{D}_R$ is contained in $\mathcal{D}$ is what I consider to be the second profound result in the solution of the dimension sequence problem, and although this theorem appears in the joint paper Arnold and Gilmer [1973], its proof is due entirely to Arnold. It is interesting to note that Arnold uses a different kind of 'special chain' of prime ideals of $R^{(n)}$ in order to establish the inclusion $\mathcal{D}_R \subseteq \mathcal{D}$. He dubs these *k*-chains*, and the concept seems to be even more powerful than Jaffard's 'special chains'. For example, Arnold has shown me a proof that the difference sequence of a ring is eventually constant which is independent of Jaffard's work; his proof is based upon the theory of k*-chains, and it is high time that I gave you the definition.

If $\mathcal{C}$ is a finite chain of prime ideals of $R^{(m)}$, then for each i between 0 and m, we let $\mathcal{C}_i$ be the chain of prime ideals of $R^{(i)}$ obtained by intersecting the members of $\mathcal{C}$ with $R^{(i)}$ (where $R^{(0)} = R$). Then $\mathcal{C}$ is a *k*-chain* if

(i) $\ell(\mathcal{C}) = k$, where ℓ denotes length,

(ii) $\ell(\mathcal{C}_i) - \ell(\mathcal{C}_{i-1}) \leqslant \ell(\mathcal{C}_0) + 1$ for $1 \leqslant i \leqslant m$,

(iii) $\ell(\mathcal{C}_i) - \ell(\mathcal{C}_{i-1}) \leqslant \ell(\mathcal{C}_{i-1}) - \ell(\mathcal{C}_{i-2})$ for $2 \leqslant i \leqslant m$,

and (iv) $1 \leqslant \ell(\mathcal{C}_i) - \ell(\mathcal{C}_{i-1})$ for $1 \leqslant i \leqslant m$.

In terms of $\mathcal{S}$, $\mathcal{C}$ is a k*-chain if $\ell(\mathcal{C}) = k$ and if $\{\ell(\mathcal{C}_0), \ell(\mathcal{C}_1), \ldots, \ell(\mathcal{C}_m)\}$ is the initial segment of an element of $\mathcal{S}$.

The following theorem is Arnold's main result concerning k*-chains.

(R10) If M is a maximal ideal of $R^{(m)}$ of height k, then M is the terminal element of a k*-chain in $R^{(m)}$.

Because of the importance that the containment $\mathcal{D}_R \subseteq \mathcal{D}$ played in the final solution of the dimension sequence problem, I should like to outline a proof of this inclusion relation that indicates how (R10) is used therein. Thus, let $\{n_i\}_0^\infty$ be in $\mathcal{D}_R$ and let $\{d_i\}_1^\infty$ be the associated difference sequence. If d is the eventual value of the sequence $\{d_i\}$, then we choose u minimal so that $d_j = d$ for each $j > u$; if $u = 1$, then $\{n_i\}$ is in $\mathcal{S}$, and hence in $\mathcal{D}$, so we assume that u is greater than 1. For $1 \leqslant i \leqslant u+1$, let $\mathcal{C}^{(i)}$ be an n_i^*-chain in $R^{(i)}$ and let s_i be the sequence

$$\{\ell(\mathcal{C}^{(i)} \cap R), \ell(\mathcal{C}^{(i)} \cap R^{(1)}), \ldots, \ell(\mathcal{C}^{(i)}) = n_i, n_i+1, n_i+2, \ldots\};$$

each sequence s_i in $\mathcal{S}$, as is the sequence

$$s_{u+1} = \{\ell(\mathcal{C}^{(u+1)} \cap R), \ldots, \ell(\mathcal{C}^{(u+1)}) = n_{u+1}, n_{u+1}+d, n_{u+1}+2d, \ldots\}.$$

Moreover, $\{n_i\}_0^\infty = \sup\{s_1, s_2, \ldots, s_{u+1}\}$, and thus $\{n_i\}$ is in $\mathcal{D}$.

The equality $\mathcal{D}_R = \mathcal{D}$ completes a solution of the dimension sequence problem for rings, but even a cursory look at the development of the dimension theory of polynomial rings will indicate that a determination of $\mathcal{D}_I$ has historically been considered to be at least as important as a determination of $\mathcal{D}_R$. The equality $\mathcal{D}_R = \mathcal{D}$ provides us with a natural, if bold, conjecture: $\mathcal{D}_I = \mathcal{D}$. Since the inclusion $\mathcal{D}_I \subseteq \mathcal{D}$ follows from the inclusion $\mathcal{D}_R \subseteq \mathcal{D}$, a proof of this last conjecture reverts to the realization problem – that is, it is necessary to show that each element of $\mathcal{D}$ can be realized as the dimension sequence of an integral domain. The construction used to solve this realization problem is a generalization of the (D+M)-construction that I had investigated in some detail in Gilmer [1969]; the generalization goes as follows: Let $\{V_i\}_{i=1}^n$ be a finite family of valuation rings on a field L such that $V_i \not\subseteq V_j$ for $i \neq j$ and assume that there exists a subfield K of L such that for each i, $V_i = K + M_i$, where M_i is the maximal ideal of V_i. For each i, let D_i be a subring of K, define J_i to be $D_i + M_i$, and let $J = \cap_{i=1}^\infty J_i$. Then J is a domain with quotient field L and the dimension sequence for J is the supremum, in the cardinal order, of the dimension sequences for $J_1, J_2, \ldots J_n$.

Since each element of $\mathcal{S}$ can be realized as the dimension sequence of a (D+M)-construction, one might reasonably expect that each element of $\mathcal{D}$ can be realized as the dimension sequence of a domain J constructed as in the preceding paragraph. This is, in fact, the case, and the next result gives a bit of information in addition to the inclusion $\mathcal{D} \subseteq \mathcal{D}_I$.

(R11) Let F be a field, let $\{X_i\}_0^\infty$ be a countably infinite set of indeterminates over F, and let $s_1, s_2, \ldots$ be elements of $\mathcal{S}$. There exists an integrally closed semi-quasi-local domain D with quotient field $F(\{X_i\}_0^\infty)$ such that $\sup\{s_1, s_2, \ldots, s_n\}$ is the dimension sequence for D.

It follows from (R11) that $\mathcal{D}_I = \mathcal{D}$, and hence we have arrived at a solution of the dimension sequence problem for integral domains. It seems desirable, however, to have an intrinsic, number-theoretic characterization of the elements of $\mathcal{D}$. Two such characterizations have been given; I shall state them below as (R12) and (R13). These results appear in a forthcoming paper (Parker [to appear]) by T. Parker; (R13) was obtained independently by Bastida and C. Pomerance.

(R12) The increasing sequence $\{a_i\}_0^\infty$ is in $\mathcal{D}$ if and only if the following two conditions are satisfied.

(i) $[a_0/1] \geq [a_1/2] \geq \ldots \geq [a_n/(n+1)] \geq \ldots,$

where brackets denote the greatest integer function.

(ii) If $b_i = [a_i/(i+1)] = b_{i+1} = [a_{i+1}/(i+2)]$, and if $r_1 = a_i - (i+1)b_i$ and $r_2 = a_{i+1} - (i+2)b_{i+1}$, then $r_1 < i$ implies that $r_2 \leq r_1$.

(R13) The increasing sequence $\{a_i\}_0^\infty$ of positive integers is in $\mathcal{D}$ if and only if $na_n \leq (n+1)a_{n-1}+1$ for each positive integer n.

I shall conclude by mentioning an auxiliary problem that Arnold and I considered in Arnold and Gilmer [1973]: Suppose K is a field and s is in $\mathcal{D}$. Under what conditions does K admit a subring with dimension sequence s? It is fairly easy to establish a certain necessary condition that K must satisfy, namely, the sequence s determines the valuative dimension d of a domain with dimension sequence s, and hence K must admit valuation rings of rank at least d. Without going into detail, it turns out that this necessary condition is also sufficient. Hence K admits a subring with dimension sequence s if and only if the transcendence degree of K over its prime subfield is (1) at least d if the characteristic of K is non-zero, or (2) at least d-1 if the characteristic of K is zero.

BIBLIOGRAPHY

ARNOLD, J.T.

[1969] On the dimension theory of overrings of an integral domain, Trans. Amer. Math. Soc. 138, 1969, 313-326.

ARNOLD, J.T. and R. GILMER

[1973] Dimension sequences for commutative rings, Bull. Amer. Math. Soc. 79, 1973, 407-409.

[to appear] The dimension sequence of a commutative ring, Amer. J. Math.

BASS, H.

[1971] Descending chains and the Krull ordinal of commutative Noetherian rings, J. Pure Appl. Algebra 1, 347-360, 1971.

BASTIDA, E.R. and R. GILMER

[1973] Overrings and divisorial ideals of rings of the form D + M, Michigan Math. J. 20, 79-95, 1973.

GILMER, R.

[1969] Two constructions of Prüfer domains, J. Reine Angew. Math. 239/240, 153-162, 1969.

[1972] *Multiplicative Ideal Theory*, Dekker, New York, 1972.

[to appear] Dimension sequences of commutative rings, Proceedings of 1973 University of Oklahoma Ring Theory Conference.

GILMER, R.

[1973] Prüfer-like conditions on the set of overrings of an integral domain, Conf. on Commutative Algebra Proceedings, 1972, Lecture Notes in Math. # 311, Springer-Verlag, New York, 1973.

JAFFARD, P.

[1960] *Théorie de la Dimension dans les Anneaux de Polynomes*, Gauthier-Villars, Paris, 1960.

KRULL, W.

[1951] Jakobsonsche Ringe, Hilbertsche Nullstellensatz, Dimensionentheorie, Math. Zeit. 54, 354-387, 1951.

[1932] Allgemeine Bewertungstheorie, J. Reine Angew. Math. 167, 160-196, 1932.

PARKER, T.

[to appear] A number-theoretic characterization of dimension sequences, Amer. J. Math.

PRÜFER, H.

[1932] Untersuchungen über die Teilbarkeitseigenschaften in Körpern, J. Reine Angew. Math. 168, 1-36, 1932.

SEIDENBERG, A.

[1953] A note on the dimension theory of rings, Pacific J. Math. 3, 505-512, 1953.

[1954] On the dimension theory of rings II, Pacific J. Math. 4, 603-614, 1954.

* * *

Department of Mathematics, Florida State University, Tallahassee, Florida, USA

Department of Mathematics, La Trobe University, Bundoora, Victoria, Australia.

DIMENSION THEORY OF POWER SERIES RINGS OVER A COMMUTATIVE RING

Robert Gilmer

This paper is a survey of some known results concerning the dimension theory of power series rings in finitely many indeterminates over a commutative ring with identity.

As in the preceding paper, all rings considered herein are assumed to be commutative and to contain an identity element. We consider power series rings in finitely many indeterminates over a ring R, and we use the symbol $R^{[m]}$ to denote the power series ring $R[[X_1,\ldots,X_m]]$. Again the questions discussed are of no interest for infinite-dimensional rings, so we make the additional assumption that all rings considered have finite (Krull) dimension.

Compared to the state of knowledge in regard to the dimension theory of polynomial rings, the theory for power series rings is still in its childhood, perhaps at a comparable stage as the dimension theory of polynomial rings after the work of A. Seidenberg. In particular, the problem of determining all 'power dimension sequences' seems premature at this point; several basic questions must be answered before success on that problem is to be anticipated. To provide a basis for my remarks on the dimension theory of power series rings, I list a few results concerning the dimension of polynomial rings; these would seem to be the only results from the polynomial theory that are pertinent to the power series theory at this stage. The notation in stating these results will be as in the previous paper.

THEOREM A (Seidenberg)

$n_0 + 1 \leq \dim R^{(1)} \leq 2n_0 + 1$; *for each* k, $\dim R^{(k)}$ *is finite.*

THEOREM B (Krull)

For a Noetherian ring R, $\dim R^{(k)} = n_0 + k$ *for all* k.

COROLLARY C

$\dim K^{(n)} = n$ *for all* n *if* K *is a field.*

COROLLARY D (Jaffard)

For a fixed proper prime ideal P *of* R, *a chain of prime ideals of* $R^{(m)}$, *each lying over* P *in* R, *contains at most* $m + 1$ *members.*

COROLLARY E (Jaffard)

$n_0 + m \leq \dim R^{(m)} \leq (n_0+1)(m+1) - 1$ *for each positive integer* m.

COROLLARY F

For a zero-dimensional ring (and in particular for a regular ring) R, $\dim R^{(m)} = m$ *for all* m.

THEOREM G (Seidenberg)

For a Prüfer domain R, $\dim R^{(k)} = n_0 + k$ *for each positive integer* k.

In considering Theorem A, there is no problem in establishing the inequality $n_0 + 1 \leq \dim R^{[1]}$; we merely examine the proof in the polynomial ring case, which goes as follows. If A is an ideal of R, then the canonical homomorphism ϕ of R onto R/A has a unique extension to a homomorphism ϕ^* of $R^{(m)}$ onto $(R/A)^{(m)}$ with $\phi^*(X_i) = X_i$ for each i between 1 and m; this is just the mapping that reduces each coefficient of an element of $R^{(m)}$ modulo A, and its kernel is the ideal $A^{(m)} = A\cdot R^{(m)}$ of $R^{(m)}$ consisting of all polynomials, each of whose coefficients belongs to the ideal A. Hence $R^{(m)}/A^{(m)}$ is isomorphic to $(R/A)^{(m)}$, and thus $A^{(m)}$ is prime in $R^{(m)}$ if A is prime in R. Therefore, if $P_0 \subset P_1 \subset \ldots \subset P_t$ is a chain of proper prime ideals of R, then $P_0^{(m)} \subset P_1^{(m)} \subset \ldots \subset P_t^{(m)} \subset P_t^{(m)} + (X_1) \subset \ldots \subset P_t^{(m)} + (X_1,\ldots,X_m)$ is a chain of proper prime ideals of $R^{(m)}$ and this establishes the inequality $n_0 + m \leqslant \dim R^{(m)}$. Similarly, there is a unique extension ϕ^{**} of ϕ to a homomorphism of $R^{[m]}$ onto $(R/A)^{[m]}$ such that $\phi^{**}(X_i) = X_i$ for each i; ϕ^{**} is defined by

$$\phi^{**}(\Sigma a_{i_1\ldots i_n} X_1^{i_1}\ldots X_n^{i_n}) = \Sigma\phi(a_{i_1\ldots i_n}) X_1^{i_1}\ldots X_n^{i_n},$$

and its kernel is the ideal $A^{[m]}$ of $R^{[m]}$ consisting of all power series, each of whose coefficients belongs to the ideal A. It is worth noting that the ideal $A^{[m]}$ may properly contain $A\cdot R^{[m]}$, the smallest ideal of $R^{[m]}$ containing A, for $A\cdot R^{[m]}$ consists of those power series f such that all coefficients of f belong to some finitely generated ideal contained in A. At any rate, $A^{[m]}$ is prime in $R^{[m]}$ if A is prime in R so we obtain the inequality $n_0 + m \leqslant \dim R^{[m]}$ as in the case of polynomial rings.

One encounters difficulty in trying to establish an upper bound on $\dim R^{[1]}$, and for good reason – it turns out that $R^{[1]}$ may be infinite-dimensional, although R is finite-dimensional. (We discuss this result later.) This means that the analogue of Corollary D fails for power series rings, and it is instructive to see why the proof of Corollary D does not generalize to $R^{[m]}$. The proof of Corollary D goes as follows. Let $P_0 \subset P_1 \subset \ldots \subset P_t$ be a chain of proper prime ideals of $R^{(m)}$ such that $P_0 \cap R = P_1 \cap R = \ldots = P_t \cap R = P$. We wish to prove that $t \leqslant m$. Since $P^{(m)}$ is a prime ideal of $R^{(m)}$ contained in P_j, we assume without loss of generality that $P_0 = P^{(m)}$. Then by means of the isomorphism $R^{(m)}/P^{(m)} \simeq (R/P)^{(m)}$, we obtain a chain

$(0) \subset P_1' \subset \ldots \subset P_t'$ of proper primes of $D^{(m)}$ (where D is the domain R/P) such that $P_i' \cap D = (0)$ for each i. Therefore each P_i' fails to meet the multiplicative system $N = D - \{0\}$ of nonzero elements of D. Consequently, each P_i' extends to a proper prime ideal P_i'' of the quotient ring $(D^{(m)})_N$ of $D^{(m)}$, and we obtain the chain

$P_0'' \subset P_1'' \subset \ldots \subset P_t''$ of proper primes of $(D^{(m)})_N$. Since $(D^{(m)})_N = (D_N)^{(m)}$, where D_N is the quotient field of D, and since $\dim (D_N)^{(m)} = m$, we conclude that $t \leqslant m$.

There are two places in the preceding proof at which one encounters difficulties in seeking to establish a power series analogue; they are:

(1) If Q is a prime ideal of $R^{[m]}$ and if $P = Q \cap R$, then Q need not contain $P^{[m]}$. Of course, Q contains $P \cdot R^{[m]}$, but $P^{[m]}$ may not be contained in the radical of $P \cdot R^{[m]}$.

(2) The inclusion $(D^{[m]})_N \subseteq D_N^{[m]}$ may be proper. This relation gives trouble in many different questions concerning power series rings – see, for example, Gilmer [1967], Rivet [1967, 1969, 1971], and Sheldon [1971] – and the equality $(D^{[m]})_N = (D_N)^{[m]}$ is certainly the exception, rather than the rule.

There might appear to be a third difficulty in seeking an analogue of Corollary D, that of the equality $\dim K^{[m]} = m$ for a field K. This happens not to be a problem, however, for in Fields [1970], D.E. Fields presents a proof, due to W. Heinzer, that $\dim R^{[m]} = n_0 + m$ for a Noetherian ring R. It is noteworthy that this proof is possible because Krull's generalized principal ideal theorem provides an alternate characterization of the height of a proper prime ideal P of a Noetherian ring: P has height at most k if and only if P is a minimal prime of an ideal with a basis of k elements.

The paper (Fields, [1970]) of Fields represents the first work devoted to the dimension theory of power series rings. The obvious starting point was the question of whether $R^{(1)}$ is finite-dimensional, and based on the considerations of Krull, Seidenberg, and P. Jaffard for polynomial rings, it seemed advisable to first look at Noetherian rings, Prüfer domains, and rings of small dimension. The answer to the question, for Noetherian rings, came fairly soon, but the case of Prüfer domains seemed to be of sufficient difficulty to indicate initial consideration of quasi-local Prüfer domains — that is, valuation rings. In this connection, Fields showed that if V is a valuation ring of rank n_0 (for valuation rings it is customary to use the word rank instead of dimension) and if V contains no nonzero idempotent proper prime ideal, then $\dim V^{[1]} = n_0 + 1$. But for a rank one nondiscrete valuation ring V with maximal ideal M, Fields was unable to determine whether the dimension of $V^{[1]}$ is finite; he showed that $\dim V^{[1]} \geq 3$ (which already indicates variance from the case of polynomial rings) and *if* $M \cdot V^{[1]}$ *is prime in* $V^{[1]}$, then $\dim V^{[1]} \geq 4$. More generally, Fields proved that if V is a rank n_0 valuation ring with k nonzero idempotent proper prime ideals, then $\dim V^{[1]} \geq n_0 + k + 1$. In Arnold and Brewer [1973], J.T. Arnold and J.W. Brewer showed that $M \cdot V^{[1]}$ is prime in $V^{[1]}$, and thus $\dim V^{[1]} \geq 4$ if V is a rank one nondiscrete valuation ring.

At this point one suspected that the power series ring in one indeterminate over a rank one nondiscrete valuation ring is infinite-dimensional, and Arnold confirmed this suspicion in his significant paper (Arnold [1973]). In Arnold [1973], Arnold introduces the notions of SFT-*ideal* and SFT-*ring* (SFT is an abbreviation for *strong finite type*); the definitions are as follows. The ideal A is an SFT-*ideal* if there is a finitely generated ideal B contained in A and a positive integer k such that $x^k \in B$ for each x in A; the ring R is an SFT-*ring* if each ideal of R is an SFT-ideal. In what, to me, is a very difficult proof, Arnold showed that $R^{[1]}$ is infinite-dimensional if R is not an SFT-ring. This theorem yields a number of interesting consequences.

(1) If P is a nonzero proper prime ideal of a valuation ring, then P is an SFT-ideal if and only if $P \neq P^2$. Thus $V[[X]]$ is infinite-dimensional if V is a valuation ring containing an idempotent nonzero proper prime ideal.

(2) Each ideal of a regular ring is equal to its radical. Hence SFT-ideals of a regular ring are finitely generated, and $R[[X]]$ is finite-dimensional, for R regular, if and only if R is Noetherian.

Arnold also used the SFT-condition to show that a one-dimensional Prüfer domain D containing no idempotent nonzero proper prime ideal (that is, *almost Dedekind domain*) need not be such that $D^{[1]}$ is finite-dimensional.

In Arnold [1973], Arnold considers finite-dimensional Prüfer domains J that are SFT-rings; he shows that $\dim J^{[1]} = \dim J + 1$ in this case, and hence for Prüfer domains the SFT-condition is equivalent to the condition that the power series ring in one indeterminate should be finite-dimensional. In analogy with the case of polynomial rings, he showed that if P is a proper prime ideal of J, then each prime ideal of $J[[X]]$ contained in $P[[X]]$ is of the form $P_1[[X]]$ for some prime P_1 of J contained in P.

That brings us to a fair summary of what is known about the dimension theory of power series rings. There are many open questions in the area; I shall mention three.

(Q1) If R is an SFT-ring, is $R^{[1]}$ finite-dimensional? Is $R^{[1]}$ an SFT-ring?

(Q2) If $R^{[1]}$ is finite-dimensional, is $R^{[2]}$ also finite-dimensional?

(Q3) if $R^{[1]}$ is finite-dimensional, is $\dim R^{[1]} = \dim R+1$?

With regard to (Q1), it is known (Arnold [1972]) that R is an SFT-ring if and only if $P^{[1]}$ is contained in the radical of $P \cdot R^{[1]}$ for each prime ideal P of R. This means that the first difficulty mentioned in connection with generalizing Corollary D to power series rings disappears for SFT-rings. Hence the first part of (Q1) has the same answer that the following question has.

(Q1)' If D is an SFT-domain and if $N = D - \{0\}$, is the domain $(D^{[1]})_N$ finite-dimensional?

The questions mentioned may be too general, and additional restrictions, such as to Prüfer domains, may initially be advisable.

BIBLIOGRAPHY

ARNOLD, J.T.

[1972] Prime ideals in power series rings, Conference on Commutative Algebra Proceedings 1972, Lecture Notes in Mathematics #311, Springer-Verlag, New York, 1972.

[1973] Power series rings over Prüfer domains, Pacific J. Math. 44, 1-11, 1973.

[1973] Krull dimension in power series rings, Trans. Amer. Math. Soc. 177, 1973, 299-304.

ARNOLD, J.T. and J.W. BREWER

[1973] On when $(D[[X]])_{P[[X]]}$ is a valuation ring, Proc. Amer. Math. Soc. 37, 326-332, 1973.

FIELDS, D.E.

[1970] Dimension theory in power series rings, Pacific J. Math. 35, 601-611, 1970.

GILMER, R.

[1967] A note on the quotient field of the domain $D[[X]]$, Proc. Amer. Math. Soc. 18, 1138-1140, 1967.

RIVET, R.

[1967] Sur le corps de fractions d'un anneau de séries formelles à coefficients dans un anneau de valuation discrète, C.R. Paris Acad. Sci. Sér. A 264, 1047-1049, 1967.

[1969] Sur les fonctions à valeurs entières, associées au corps des fractions d'un anneau de séries formelles à coefficients dans un anneau de valuation discrète, C.R. Paris Acad. Sci. Sér. A 268, 1455-1457, 1969.

[1971] Famille d'anneaux locaux henseliens dominés par $Cf(A)[[X]]$, défines par des fonctions pseudo-concaves, C.R. Paris Acad. Sci. Sér. A 272, 369-371, 1971.

SHELDON, P.

[1971] How changing $D[[X]]$ changes its quotient field, Trans. Amer. Math. Soc. 159, 223-244, 1971.

AXIOMATIC CLASSES IN PROPOSITIONAL MODAL LOGIC

R.I. Goldblatt and S.K. Thomason†

In his review (Kaplan [1966]) of the article in which Kripke first proposed his relational semantics for modal logic, David Kaplan posed the question: which properties of a binary relation are expressible by formulas of propositional modal logic? A class of Kripke frames is said to be *modal-axiomatic* if it comprises exactly the frames on which every one of some set of formulas of propositional modal logic is valid. This work is addressed to the problem, suggested by Kaplan's question, of characterizing the modal-axiomatic classes of Kripke frames. In §1 we obtain such a characterization, in terms of closure under certain constructions. In §2 we show that, in the case of classes closed under elementary equivalence, much simpler constructions suffice.

It is assumed that the reader is familiar with the Kripke semantics for modal logic (see e.g. Cresswell [1975] or Segerberg [1971]).

A *first-order frame* is a triple $\langle W,R,P \rangle$, where W is a non-empty set, R is a binary relation on W, and P is a non-empty collection of subsets of W closed under the Boolean operations and the unary operation M_R defined by

†The work of the second-named author was supported in part by the Canada Council and the National Research Council of Canada.

$$M_R X = \{y \in W \mid (\exists x)(yRx \ \& \ x \in X)\}.$$

A *Kripke frame* $\langle W,R \rangle$ may be identified with the first-order frame $\langle W,R,2^W \rangle$. The notion "α is valid on $\langle W,R,P \rangle$" is defined like "α is valid on $\langle W,R \rangle$", except that only valuations V such that $V(p) \in P$, for all propositional variables p, are considered. A *modal algebra* is an algebra $\mathfrak{a} = \langle A;0,1,\cup,\cap,',M \rangle$ such that $\langle A;0,1,\cup,\cap,' \rangle$ is a Boolean algebra and M is a unary operation satisfying $M0 = 0$ and $M(a \cup b) = Ma \cup Mb$. If $\langle W,R,P \rangle$ is a frame then $\langle W,R,P \rangle^+ = \langle P;\emptyset,W,\cup,\cap,',M_R \rangle$ is a modal algebra. To each formula α of propositional modal logic there corresponds a modal algebra polynomial identity e_α such that α is valid on a frame $\langle W,R,P \rangle$ $(\langle W,R,P \rangle \models \alpha)$ if and only if e_α holds in $\langle W,R,P \rangle^+$. Conversely, to each identity e there corresponds a formula α_e, and $\langle W,R,P \rangle \models \alpha_e$ if and only if e holds in $\langle W,R,P \rangle^+$.

§1. We shall first describe in intuitive terms the most complex of the constructions with which we shall be concerned. The reader will recall that in a Kripke frame $\langle W,R \rangle$, W is meant to represent the collection of all *possible worlds*, and R the *accessibility* relation between worlds. Then each subset of W represents a possible extension of a propositional variable, i.e. a *proposition*. A *state of affairs* is a specification for each proposition whether or not it obtains, i.e. an ultrafilter in $\langle W,R \rangle^+$. One state of affairs is a *possible alternative* to another if every proposition true in the first is possible in the second. Thus the Kripke frame $\langle U,S \rangle$, where U is the set of all ultrafilters in $\langle W,R \rangle^+$ and $uSv \leftrightarrow (\forall X \subseteq W)(X \in v \rightarrow M_R X \in u)$, may be interpreted as the frame of all states of affairs relative to the frame $\langle W,R \rangle$ of possible worlds.

In general, not every propositional form valid in $\langle W,R\rangle$ will be valid in $\langle U,S\rangle$, because some propositions represented in $\langle U,S\rangle$ will not be represented in $\langle W,R\rangle$. More precisely, a proposition is represented in $\langle U,S\rangle$ by a set $Y \subseteq U$; the same proposition will be represented in $\langle W,R\rangle$ by $X \subseteq W$ if and only if $(\forall u{\in}U)(X{\in}u \Leftrightarrow u{\in}Y)$, and in general no such X need exist. But as we shall see, it is possible to generalize the construction and impose fairly simple restrictions which guarantee that validity *is* preserved in passing from $\langle W,R\rangle$ to $\langle U,S\rangle$.

Given a Kripke frame $\langle W,R\rangle$, a first-order frame $\langle W,R,P\rangle$ represents a decision that for current purposes the only propositions which matter are those represented by sets $X \in P$. Then a state of affairs relative to $\langle W,R,P\rangle$ is represented by an ultrafilter in $\langle W,R,P\rangle^{+}$. One such state of affairs is a possible alternative to another if every material proposition true in the one is possible in the other, so we define S by

(1) $$uSv \Leftrightarrow (\forall X{\in}P)(X{\in}v \Rightarrow M_R X{\in}u).$$

Besides deciding that only certain propositions are material, we may decide that only certain states of affairs are conceivable; such a decision is represented by a set U of ultrafilters in $\langle W,R,P\rangle^{+}$. Now under what conditions can we expect that every propositional form valid in $\langle W,R\rangle$ is valid in $\langle U,S\rangle$?

As before, it will be necessary that no "new" propositions be represented in $\langle U,S\rangle$; thus we require

(2) $$(\forall Y{\subseteq}U)(\exists X{\in}P)(\forall u{\in}U)(u{\in}Y \Leftrightarrow X{\in}u).$$

Moreover, when we admit a state of affairs u as conceivable we must also, for each proposition possible in u, admit as conceivable a possible alternative v to u in which the proposition is true, i.e. we require

(3) $(\forall u\in U)(\forall X\in P)(M_R X\in u \Rightarrow (\exists v\in U)(uSv \ \& \ X\in v))$.

DEFINITION 1

$\langle U,S\rangle$ is a *states-of-affairs frame based on* $\langle W,R\rangle$, or more briefly $\langle U,S\rangle$ is *SA-based on* $\langle W,R\rangle$, if there is a first-order frame $\langle W,R,P\rangle$ such that U is a set of ultrafilters in $\langle W,R,P\rangle^+$ and (1), (2), and (3) are satisfied. A class K of Kripke frames is *closed under SA-constructions* if every frame SA-based on a member of K is a member of K.

If L is a class of algebras, then HS(L) is the class of all homomorphic images of subalgebras of members of L; $HS(\mathfrak{a}) = HS(\{\mathfrak{a}\})$.

LEMMA 1

$\langle U,S\rangle$ is isomorphic to a frame SA-based on $\langle W,R\rangle$ if and only if $\langle U,S\rangle^+ \in HS(\langle W,R\rangle^+)$.

Proof. If $\langle U,S\rangle$ is SA-based on $\langle W,R\rangle$, let $\langle W,R,P\rangle$ be as in Definition 1. Then $\langle W,R,P\rangle^+$ is a subalgebra of $\langle W,R\rangle^+$. Define $f : \langle W,R,P\rangle^+ \to \langle U,S\rangle^+$ by $f(X) = \{u \mid X\in u\}$. Clearly f is a Boolean algebra homomorphism, and because of (2) f is onto. By (1) and (3), $M_R X \in u \Leftrightarrow (\exists v)(uSv \ \& \ X\in v)$; hence $f(M_R X) =$ $\{u \mid M_R X \in u\} = \{u \mid (\exists v)(uSv \ \& \ X\in v)\} = M_S\{v \mid X\in v\} = M_S f(X)$. Thus f is a homomorphism from $\langle W,R,P\rangle^+$ onto $\langle U,S\rangle^+$, and $\langle U,S\rangle^+ \in HS(\langle W,R\rangle^+)$.

Conversely, suppose $\langle U,S\rangle^+ \in HS(\langle W,R\rangle^+)$. Every subalgebra of $\langle W,R\rangle^+$ is of the form $\langle W,R,P\rangle^+$, so let $h : \langle W,R,P\rangle^+ \to \langle U,S\rangle^+$ be a homomorphism onto. For $u \in U$, let $g(u) = \{X{\in}P \mid u \in h(X)\}$, and let $U' = \{g(u) \mid u{\in}U\}$. Then U' is a set of ultrafilters in $\langle W,R,P\rangle^+$. Define S' so that $\langle U',S'\rangle$ satisfies (1). We claim that $\langle U',S'\rangle$ is SA-based on $\langle W,R\rangle$ and that g is an isomorphism from $\langle U,S\rangle$ to $\langle U',S'\rangle$.

Since h is onto there is, for each $u \in U$, an $X \in P$ with $h(X) = \{u\}$. Then $X \in g(u)$, but $X \notin g(v)$ when $v \neq u$. Thus g is one-to-one, and obviously g is onto. We have

$$\begin{aligned} u\ S\ v &\Rightarrow (\forall X{\in}P)(v \in h(X) \Rightarrow u \in M_S h(X) = h(M_R X)) \\ &\Leftrightarrow (\forall X{\in}P)(X \in g(v) \Rightarrow M_R X \in g(u)) \\ &\Leftrightarrow g(u)\ S'\ g(v). \end{aligned}$$

Also, if $(\forall X{\in}P)(v \in h(X) \Rightarrow u \in M_S h(X))$ then fixing X so that $h(X) = \{v\}$ yields $u \in M_S\{v\}$, i.e. $u\ S\ v$. Hence $u\ S\ v \Leftrightarrow g(u)\ S'\ g(v)$, and g is an isomorphism as claimed.

If $Y' \subseteq U'$, let $Y = g^{-1}[Y'] \subseteq U$ and choose $X \in P$ such that $h(X) = Y$. For any $u' \in U'$ we have $u' \in Y' \Leftrightarrow g^{-1}(u') \in Y = h(X) \Leftrightarrow X \in g(g^{-1}(u')) = u'$. Thus $\langle U',S'\rangle$ satisfies (2). Suppose $u' \in U'$, $X \in P$, and $M_R X \in u'$. Let $u = g^{-1}(u')$. Since $u' = g(u) = \{Y{\in}P \mid u \in h(Y)\}$, we have $u \in h(M_R X) = M_S h(X)$, i.e. $u\ S\ v$ holds for some $v \in h(X)$. Then $u'\ S'\ g(v)$ and $X \in g(v)$. Hence $\langle U',S'\rangle$ satisfies (3) and the proof is complete.

COROLLARY 2

If $\langle U,S\rangle$ is SA-based on $\langle W,R\rangle$ and $\langle W,R\rangle \models \alpha$ then $\langle U,S\rangle \models \alpha$.

Proof. Every polynomial identity holding in $\mathfrak{a}$ holds in any element of $HS(\mathfrak{a})$.

THEOREM 3

A class of Kripke frames is modal-axiomatic if and only if it is closed under isomorphism, non-trivial disjoint unions (i.e. the index set is non-empty), and SA-constructions.

Proof. Every modal-axiomatic class is closed under isomorphism and non-trivial disjoint union by standard results (Segerberg [1971], Vol. I), and under SA-constructions by Corollary 2.

Conversely, suppose K is a non-empty class of Kripke frames satisfying the closure conditions of the theorem. (The case $K = \emptyset$ is trivial.) Let $K^+ = \{\mathfrak{a} \mid (\exists \langle W,R\rangle \in K)(\mathfrak{a} \cong \langle W,R\rangle^+)\}$. Since $\Pi(\langle W_i,R_i\rangle^+ \mid i\in I) \cong (\Sigma(\langle W_i,R_i\rangle \mid i\in I))^+$ by the obvious map (where Π denotes direct product, Σ disjoint union), K^+ is closed under non-trivial direct products. Hence $HS(K^+) = HSP(K^+)$ is an equational class (Grätzer [1968], §§23,26), say $\mathfrak{a} \in HS(K^+) \Leftrightarrow (\forall e\in\Sigma)(e$ holds in $\mathfrak{a})$.

Then $\langle W,R\rangle \in K \Rightarrow \langle W,R\rangle^+ \in K^+ \Rightarrow (\forall e\in\Sigma)(\langle W,R\rangle \models \alpha_e)$. Conversely, if $(\forall e\in\Sigma)(\langle W,R\rangle \models \alpha_e)$ then $\langle W,R\rangle^+ \in HS(K^+)$, i.e. $\langle W,R\rangle^+ \in HS(\langle W_1,R_1\rangle^+)$ for some $\langle W_1,R_1\rangle \in K$. By Lemma 1, $\langle W,R\rangle$ is isomorphic to a frame SA-based on $\langle W_1,R_1\rangle$; hence $\langle W,R\rangle \in K$. Thus $K = \{\langle W,R\rangle \mid (\forall e\in\Sigma)(\langle W,R\rangle \models \alpha_e)\}$ and K is modal-axiomatic.

§2. A frame $\langle W,R,P\rangle$ is *replete* if, for every ultrafilter u in $\langle W,R,P\rangle^+$,

(4) $\qquad \cap u \neq \emptyset, \qquad$ and

(5) $\qquad \cap(M_R X \mid X \in u) \subseteq M_R(\cap u).$

If $\langle W,R,P \rangle$ satisfies (5) and

(4') $\qquad (\exists w \in W)(\cap u = \{w\}),$

then $\langle W,R,P \rangle$ is *descriptive*. The notion of replete frame, and the following lemma, are essentially due to Fine (Fine [to appear]).

LEMMA 4

Given any $\langle W,R,P \rangle$ there is a replete $\langle W',R',P' \rangle$ such that $\langle W',R',P' \rangle^+ \cong \langle W,R,P \rangle^+$ and $\langle W',R' \rangle \approx \langle W,R \rangle$ (where $\approx$ denotes elementary equivalence).

Proof. Let $P = \{X_i \mid i \in I\}$, and set $\mathfrak{A} = \langle W,R,\ldots,X_i,\ldots \rangle_{i \in I}$, a structure for the first-order language with equality, a binary relation symbol p, and unary relation symbols q_i $(i \in I)$. Let $\mathfrak{A}' = \langle W',R',\ldots,X_i',\ldots \rangle_{i \in I}$ be a 2-saturated elementary extension of $\mathfrak{A}$ [Bell, Slomson [1969], Ch.11], and let $P' = \{X_i' \mid i \in I\}$. Because $\mathfrak{A}' \approx \mathfrak{A}$, we have $\langle W',R' \rangle \approx \langle W,R \rangle$ and $\langle W',R',P' \rangle^+ \cong \langle W,R,P \rangle^+$ by the map $X_i' \to X_i$. If u is an ultrafilter in $\langle W',R',P' \rangle^+$ then $\{q_i x \mid X_i \in u\}$ is finitely satisfiable in $\mathfrak{A}'$ and hence, since $\mathfrak{A}'$ is 2-saturated, simultaneously satisfiable in $\mathfrak{A}'$; i.e. $\cap u \neq \emptyset$. Similarly if $w \in \cap(M_R X \mid X \in u)$ then $\{p\underline{w}x \,\&\, q_i x \mid X_i \in u\}$ (where $\underline{w}$ is a name of w) is finitely, hence simultaneously, satisfiable in $\mathfrak{A}'$; i.e. $w \in M_R(\cap u)$. Thus $\langle W',R',P' \rangle$ is replete.

A Kripke frame $\langle W',R'\rangle$ is said to be a *p-morphic image* of $\langle W,R\rangle$ if there is a function f mapping W onto W' such that, for all $x \in W$ and $x' \in W'$,

$$f(x)\ R'\ x' \Leftrightarrow (\exists y)(xRy \ \&\ f(y) = x');$$

$\langle W',R'\rangle$ is a *subframe* of $\langle W,R\rangle$ if $W' \subseteq W$, $R' = R|W'$, and, for all $x' \in W'$ and $x \in W$,

$$x'\ R\ x \Rightarrow x \in W'.$$

In either case we have $\langle W,R\rangle \models \alpha \Rightarrow \langle W',R'\rangle \models \alpha$ (Segerberg [1971], Vol. I).

LEMMA 5

If $\langle W',R',P'\rangle$ is replete then there is a descriptive $\langle W'',R'',P''\rangle$ such that $\langle W'',R''\rangle$ is a p-morphic image of $\langle W',R'\rangle$ and $\langle W'',R'',P''\rangle^+ \cong \langle W',R',P'\rangle^+$.

Proof. Define $\sim$ on W' by $x \sim y \Leftrightarrow (\forall X \in P')(x \in X \Leftrightarrow y \in X)$, and set $W'' = W'/\sim$. Let $[x]\ R''\ [y] \Leftrightarrow (\forall X \in P')(y \in X \Rightarrow x \in M_{R'}X)$, and $P'' = \{\{[x] | x \in X\} \mid X \in P'\}$. Then $\langle W'',R'',P''\rangle$ is descriptive, $x \to [x]$ is a p-morphism from $\langle W',R'\rangle$ onto $\langle W'',R''\rangle$, and $X \to \{[x] | x \in X\}$ is an isomorphism from $\langle W',R',P'\rangle^+$ onto $\langle W'',R'',P''\rangle^+$.

COROLLARY 6

If $\mathfrak{a}$ is a subalgebra of $\langle W,R\rangle^+$ then there is a descriptive $\langle W'',R'',P''\rangle$ such that $\langle W'',R'',P''\rangle^+ \cong \mathfrak{a}$ and $\langle W'',R''\rangle$ is a p-morphic image of some $\langle W',R'\rangle \approx \langle W,R\rangle$.

Proof. Every subalgebra of $\langle W,R\rangle^+$ is of the form

$\langle W,R,P\rangle^+$. Apply Lemmas 4 and 5.

Given a Kripke frame $\langle W,R\rangle$, let $\langle W,R\rangle^* = \langle U,S\rangle$, where U is the set of all ultrafilters in $\langle W,R\rangle^+$ and $u \, S \, v \Leftrightarrow (\forall X \subseteq W)(X \in v \Rightarrow M_R X \in u)$. The next proposition follows from the duality (Goldblatt [1974], §10) between the categories of modal algebras (with homomorphisms) and descriptive frames (with appropriately defined morphisms).

PROPOSITION 7

(i) If $\langle W,R\rangle^* \models \alpha$ then $\langle W,R\rangle \models \alpha$.

(ii) If $\langle W,R,P\rangle$ is descriptive and $\langle W',R'\rangle^+$ is a homomorphic image of $\langle W,R,P\rangle^+$, then $\langle W',R'\rangle^*$ is isomorphic to a subframe of $\langle W,R\rangle$.

THEOREM 8

Let K *be a class of Kripke frames closed under elementary equivalence. Then* K *is modal-axiomatic if and only if* K *is closed under subframes, p-morphic images, and non-trivial disjoint unions, and whenever* $\langle W,R\rangle^* \in K$ *then* $\langle W,R\rangle \in K$.

Proof. The necessity of the conditions follows from Proposition 7 (i) and previous remarks. Conversely, suppose K satisfies the conditions. As in the proof of Theorem 3, it suffices to show that if $\langle W_1,R_1\rangle^+ \in HS(\langle W,R\rangle^+)$ and $\langle W,R\rangle \in K$, then $\langle W_1,R_1\rangle \in K$. Suppose then that $\langle W,R\rangle \in K$ and $\langle W_1,R_1\rangle^+$ is a homomorphic image of a subalgebra $\mathfrak{a}$ of $\langle W,R\rangle^+$. By Corollary 6 there is a descriptive frame $\langle W'',R'',P''\rangle$ with $\langle W'',R'',P''\rangle^+ \cong \mathfrak{a}$ and

$\langle W'',R''\rangle \in K$. Then $\langle W_1,R_1\rangle^+$ is a homomorphic image of $\langle W'',R'',P''\rangle^+$. By Proposition 7 (ii), $\langle W_1,R_1\rangle^*$ is isomorphic to a subframe of $\langle W'',R''\rangle$, whence $\langle W_1,R_1\rangle^* \in K$ and $\langle W_1,R_1\rangle \in K$.

It is not difficult to show that none of the four closure conditions of Theorem 8 can be deleted, even if the hypothesis "K is closed under $\approx$" is strengthened to "K is elementary". If K is the class of all $\langle W,R\rangle$ satisfying $(\forall x)(\exists y)(yRx)$, then clearly K is closed under p-morphic images and non-trivial disjoint unions and is not closed under subframes; also if $\langle W,R\rangle^* = \langle U,S\rangle \in K$ and $w \in W$, there exists $u \in U$ such that $u\ S\ \{X\subseteq W | w\in X\}$, i.e. $(\forall X\subseteq W)(w \in X \Rightarrow M_R X \in u)$, whence $M_R\{w\} \in u$ and $M_R\{w\} \neq \emptyset$; thus $\langle W,R\rangle^* \in K \Rightarrow \langle W,R\rangle \in K$. The class of all irreflexive frames satisfies all the closure conditions except that for p-morphic images, and the class of all one-element frames satisfies all the conditions except for that for non-trivial disjoint unions. Finally, the class K of all frames satisfying $(\forall x)(\exists y)(xRy \;\&\; yRy)$ is closed under subframes, p-morphic images, and non-trivial disjoint unions, but $\langle \{0,1,\ldots\}, <\rangle^* \in K$ and $\langle \{0,1,\ldots\}, <\rangle \notin K$ (Thomason [to appear]).

Finally, it may be noted that the closure conditions of Theorem 3 seem to require for their expression the notion of first-order frame (or, equivalently, of modal algebra), while the conditions of Theorem 8 are expressible strictly in terms of Kripke frames. (An ultrafilter in $\langle W,R\rangle^+$ is just an ultrafilter over W, so the reference, in the definition of $\langle W,R\rangle^*$, to a modal algebra is eliminable.) A characterization of the latter sort is of course to be preferred. It seems unlikely that the restriction in Theorem 8, to classes closed under elementary equivalence, can be dropped.

BIBLIOGRAPHY

BELL, J.L. and A.B. SLOMSON

[69] Models and Ultraproducts, North-Holland, 1969.

CRESSWELL, M.J.

[75] Frames and models in modal logic, these Proceedings.

FINE, K.

[to appear] Some connections between elementary and modal logic, Proceedings of the Third Scandinavian Symposium on Logic, North-Holland.

GOLDBLATT, R.I.

[74] Metamathematics of Modal Logic, Ph.D. Thesis, Victoria University of Wellington, 1974.

GRÄTZER, G.

[68] Universal Algebra, van Nostrand, 1968.

KRIPKE, S.A.

[66] *Semantical analysis of modal logic I*, Review by D. Kaplan, in J. Symbolic Logic 31 (1966), 120-122.

SEGERBERG, K.

[71] An Essay in Classical Modal Logic, Uppsala Universitet, 1971.

THOMASON, S.K.

[to appear] Categories of frames for modal logic.

* * *

Department of Mathematics, Victoria University of Wellington, Wellington, New Zealand

Department of Mathematics, Simon Fraser University, Burnaby, Canada

NILPOTENT ACTIONS ON NILPOTENT GROUPS

Peter Hilton

0. *Introduction*

The application of localization theory to topology (Hilton, Mislin, Roitberg [1973], Sullivan [1970]) has led to the study of nilpotent actions of nilpotent groups on commutative groups. For the appropriate setting for localization theory in topology is that of nilpotent spaces, these being characterized as connected spaces X such that the fundamental group $\pi_1 X$ is nilpotent and operates nilpotently on the higher homotopy groups $\pi_n X$, $n \geq 2$. Such a (purely algebraic) study of localization was undertaken in Hilton [1974], based on results obtained in an earlier work (Hilton [1973]).

Although there is no immediately evident topological application*, it seems natural to extend the study to nilpotent actions of nilpotent groups on nilpotent groups, and this paper is devoted to this extension. We begin, in Section 1, by completing, and elaborating, the subgroup theorems given in[1] Hilton [1974]. Not all these theorems are applied in the sequel, but it seemed sensible to gather them together in one place. Our arguments are based largely on the fundamental criterion for recognizing the P-localization of a

[1]Theorem 1.2 was announced while Hilton [1974] was in proof, so that no demonstration of parts (ii), (iii) of this theorem was given in Hilton [1974].

*(Added in proof) There is! Such actions arise in the study of *nilpotent fibrations*.

nilpotent group, where P is a family of primes (see Theorem 4.2 of Hilton [1973]), together with a theorem due to N. Blackburn [1965].

In Section 2 we consider the action of a group Q on a group N and define (2.1) the lower central Q-series of N, with respect to this action. We can then introduce the notion of a Q-nilpotent group, any such group N being automatically nilpotent. The definitions of lower central Q-series and Q-nilpotency generalize those given in Hilton [1974] for commutative groups N. Now if N is a nilpotent (but not necessarily Q-nilpotent) Q-group, then there is a unique action of Q on the P-localization N_P such that $e : N \to N_P$ is a Q-map, and Theorem 2.8 asserts that localization commutes with passage through the lower central Q-series.

In Section 3 we specialize further by supposing that not only is N Q-nilpotent but that Q is also nilpotent. We show that, given any exact sequence

(0.1) $$N \rightarrowtail G \twoheadrightarrow Q,$$

compatible with the Q-action in a certain sense (Definition 2.1) which forces G to be nilpotent, then (0.1) embeds, essentially uniquely, in a diagram

(0.2) $$\begin{array}{ccccc} N & \rightarrowtail & G & \twoheadrightarrow & Q \\ {\scriptstyle e}\downarrow & & \downarrow{\scriptstyle \phi} & & \| \\ N_P & \rightarrowtail & H & \twoheadrightarrow & Q \end{array}$$

and that H is also nilpotent. We also show (Theorem 3.3) that there is a one-one correspondence between nilpotent actions of Q on N_P and nilpotent actions of Q_P on N_P.

In Section 4, we study the subgroup N^Q of N consisting of elements fixed under Q. Our main results here are that $(N^Q)_P = N_P^Q$, provided that Q is finitely generated (but with no other conditions on Q or on the action); and that $N_P^{Q_P} = N_P^Q$, provided that Q is nilpotent and acts nilpotently on N_P.

Section 5 is an appendix giving some further remarks on the subgroup $[H,K]$ of G, where H, K are themselves subgroups of G.

We make four small changes of notation and terminology compared with Hilton [1974]. First, our notations, $\Gamma_Q^i N$, for the terms of the lower central Q-series of N, and N^Q for the fixed subgroup of N, do not contain a symbol for the action ω itself of Q on N; this seems reasonable when different actions are not under discussion – or only rarely. Second, we use $A(Q,N)$ in Section 3 for the set of *nilpotent* actions of Q on N, rather than $A_\nu(Q,N)$ as in Hilton [1974], since only nilpotent actions are in question in that section. Third, we index the terms of the lower central Q-series starting with 0 instead of with 1 as in Hilton [1973, 1974] (we thus also depart from the convention of, say, Baumslag [1971] which we followed in Hilton [1973, 1974]). The new convention has the merit that $\mathrm{nil}_Q N \leq c$ if and only if $\Gamma_Q^c N = \{1\}$; and that, for N commutative, $\Gamma_Q^i N$ is the image of N under $(IQ)^i$, rather than $(IQ)^{i-1}$, where IQ is the augmentation ideal of $\mathbb{Z}Q$. Fourth, we speak of a 'P-bijection' instead of a 'P-isomorphism' in order to emphasize that the notion is just that of a morphism which is P-injective and P-surjective, and that no question of invertibility arises.

With these exceptions, our conventions are precisely those of Hilton [1973, 1974]. In particular, P is a family of primes and the phrase $n \in P'$ means that n is a product of primes not in the family P.

1. *Subgroup theorems*

Throughout this section we assume G *nilpotent and* H, K *subgroups of* G. We will need the following theorem due to Blackburn [1965].

THEOREM 1.1

For each prime p *and each positive integer* c, *there exists a positive integer* $f = f(p,c)$, *such that, if* $\text{nil}\ G \leqslant c$, *then every product of* p^{m+f}th *powers of elements of* G *is itself a* p^mth *power.*

We now prove

THEOREM 1.2

(i) $(H \cap K)_P = H_P \cap K_P$;

(ii) $(HK)_P = H_P K_P$;

(iii) $[H,K]_P = [H_P,K_P]$.

Proof. Assertion (i) was proved in Hilton [1974]. To prove (ii) we observe that $e : G \to G_P$ respects subgroups, so that $H_P \subseteq (HK)_P$, $K_P \subseteq (HK)_P$, whence $H_P K_P \subseteq (HK)_P$. Thus, since $e(HK) \subseteq H_P K_P$, it only remains to show that $H_P K_P$ is P-local. However, it further suffices to show that $H_P K_P$ has p^{th} roots for each prime $p \in P'$, since uniqueness is obvious. Finally, this follows immediately from Blackburn's Theorem, since H_P, K_P each have p^{th} roots.

We now prove (iii). It was shown in Hilton [1974] that $e_0 : [H,K] \to [H_P,K_P]$, obtained by restricting $e : G \to G_P$, is P-bijective, so we have only to show that $[H_P,K_P]$ is P-local. Again, it is plain that we have only to show that $[H_P,K_P]$ has p^{th}

roots for $p \in P'$. In fact, we prove the following stronger result, which will also be used in the sequel.

THEOREM 1.3

If K *has* p^{th} *roots, for some prime* p*, then so has* $[H,K]$.

Proof. We fix K and allow H to vary over the subgroups of G. We define the Γ_K*-series* of H by the rule

$$\Gamma_K^0 H = H, \ \Gamma_K^{i+1} H = [\Gamma_K^i H, K], \ i \geqslant 0,$$

and write $\mathrm{nil}_K H \leqslant c$ if $\Gamma_K^c H = \{1\}$. Plainly $\mathrm{nil}_K H \leqslant \mathrm{nil}\, G$ and we prove Theorem 1.3 by induction on $\mathrm{nil}_K H$, the assertion being trivial if $\mathrm{nil}_K H \leqslant 1$. Now if $[a,b] = a^{-1}b^{-1}ab$, then

$$(1.1) \qquad [a,bc] = [a,c][a,b]^c = [a,c][a,b][[a,b],c].$$

Thus an easy induction tells us that, if $a \in H$, $b \in K$, then

$$(1.2) \qquad [a,b^n] = [a,b]^n u_n, \ u_n \in [[H,K],K].$$

Now assume that K has p^{th} roots and set $n = p^{f+1}$, where f is as in Blackburn's Theorem. We have, for $a \in H$, $k \in K$, that $k = b^n$, $b \in K$, so that

$$(1.-) \qquad [a,k] = [a,b]^n u, \ u \in [[H,K],K].$$

Now it is obvious that $\mathrm{nil}_K[H,K] < \mathrm{nil}_K H$. Thus, by our inductive hypothesis, $[[H,K],K]$ has p^{th} roots, and hence n^{th} roots. Thus

$$(1.4) \qquad [a,k] = [a,b]^n v^n, \ v \in [[H,K],K].$$

We now observe that

$$(1.5) \qquad [[H,K],K] \subseteq [H,K].$$

To see this, it suffices to show that $[w,c] \in [H,K]$, where $w \in [H,K]$, $c \in K$. It thus further suffices to show that $c^{-1}wc \in [H,K]$, and so, finally, to show that $c^{-1}[r,s]c \in [H,K]$, where $r \in H$, $s,c \in K$. But

$$c^{-1}[r,s]c = [r,s]^c = [r,c]^{-1}[r,sc], \text{ by } (1.1),$$

so that (1.5) is proved. It follows that $v \in [H,K]$ in (1.4), and thus (1.4) allows us to infer immediately that every element of $[H,K]$ is a product of nth powers of elements of $[H,K]$, so that, by Blackburn's Theorem, $[H,K]$ has p^{th} roots. This proves Theorem 1.3 and completes the proof of Theorem 1.2 (iii).

COROLLARY 1.4

If K *is a normal* P-*local subgroup of* G, *then* $[H,K]$ *is* P-*local.*

THEOREM 1.5

Let $Z(H)$ *be the centralizer of* H *in* G. *Then* e *sends* $Z(H)$ *to* $Z(H_P)$ *and* $Z(H_P)$ *is* P-*local, so that*

$$Z(H)_P \subseteq Z(H_P).$$

Moreover, $Z(H)_P = Z(H_P)$ *if* H *is finitely generated.*

Proof. The argument is a trivial generalization of that given in Hilton [1973] for the case $H = G$.

2. *Nilpotent* Q-*groups*

Now suppose that N is a group upon which the group Q acts. We say that N is a Q-*group*, and we define the *lower central* Q-*series* of N by the rule

$$(2.1)\qquad \Gamma_Q^0 N = N,\ \Gamma_Q^{i+1}N = gp(\overline{x}aba^{-1}b^{-1}),\ x \in Q,\ a \in \Gamma_Q^i N,\ b \in N,\ i \geqslant 0.$$

We say that N is Q-*nilpotent of class* $\leqslant c$, and write $nil_Q \leqslant c$, if $\Gamma_Q^c N = \{1\}$. Notice that, if N is commutative, then Definition (2.1) agrees with the definition given in Section 2 of Hilton [1974].

PROPOSITION 2.1

$\Gamma_Q^i N$ is a normal Q-subgroup of N.

Proof. We first show that $\Gamma_Q^i N$ is closed under the action of Q. Indeed, we prove this by induction, since

$$y(\overline{x}aba^{-1}b^{-1}) = (yxy^{-1}\cdot ya)(yb\overline{ya}^{-1}yb),\ x,y \in Q,\ a \in \Gamma_Q^i N,\ b \in N.$$

To show that $\Gamma_Q^i N$ is normal in N, we again use induction on i and the formula

$$c\overline{x}aba^{-1}b^{-1}c^{-1} = \overline{x}a'b'a'^{-1}b'^{-1},$$

where $a' = \overline{x^{-1}c}\ a\ \overline{x^{-1}c}^{\,-1}$, $b' = cb\overline{x^{-1}c}^{\,-1}$, $x \in Q$, $a \in \Gamma_Q^i N$, $b,c \in N$.

COROLLARY 2.2

$\Gamma_Q^{i+1}N$ is a normal Q-subgroup of $\Gamma_Q^i N$.

PROPOSITION 2.3

Let $\overline{\Gamma}_Q^i N$ be defined by

$$(2.2)\qquad \overline{\Gamma}_Q^0 N = N,\ \overline{\Gamma}_Q^{i+1}N = gp([N,\overline{\Gamma}_Q^i N],\overline{x}aa^{-1}),\ x \in Q,\ a \in \overline{\Gamma}_Q^i N,\ i \geqslant 0.$$

Then

$$\overline{\Gamma}_Q^i N = \Gamma_Q^i N.$$

Proof. We again argue by induction on i, the assertion being trivial for $i = 0$. Thus we assume that $\overline{\Gamma}^i_Q N = \Gamma^i_Q N$ and prove that $\overline{\Gamma}^{i+1}_Q N = \Gamma^{i+1}_Q N$. Since $\Gamma^{i+1}_Q N$ contains all commutators $aba^{-1}b^{-1}$, $a \in \Gamma^i_Q N$, $b \in N$, and all elements $\overline{xa}a^{-1}$, $x \in Q$, $a \in \Gamma^i_Q N$, it is plain that $\overline{\Gamma}^{i+1}_Q n \subseteq \Gamma^{i+1}_Q N$.

Conversely,

$$\overline{xaba^{-1}b^{-1}} = \overline{xa}a^{-1}(aba^{-1}b^{-1}), \; x \in Q, \; a \in \Gamma^i_Q N, \; b \in N,$$

so that $\Gamma^{i+1}_Q N \subseteq \overline{\Gamma}^{i+1}_Q N$.

Plainly (using Proposition 2.3), writing $\Gamma^i N$ for the ith term of the lower central series of the group N, starting with $\Gamma^0 N = N$, we have

$$(2.3) \qquad \Gamma^i N \subseteq \Gamma^i_Q N,$$

so that we have the following corollary.

COROLLARY 2.4

$\Gamma^i_Q N / \Gamma^{i+1}_Q N$ *is a trivial* Q-*group in the centre of* $N/\Gamma^{i+1}_Q N$, $i \geqslant 0$.

Now form the semidirect product G of N and Q; we thus have a short exact sequence with right splitting σ,

$$(2.4) \qquad N \overset{\mu}{\rightarrowtail} G \underset{\underset{\sigma}{\longleftarrow}}{\overset{\varepsilon}{\twoheadrightarrow}} Q, \qquad \varepsilon\sigma = 1,$$

where $\mu b = (b,1)$, $\varepsilon(b,x) = x$, $\sigma x = (1,x)$, $b \in N$, $x \in Q$. Moreover the action of Q on N given by

$$(2.5) \qquad b \longmapsto \mu^{-1}(\sigma x \mu \overline{b} \sigma x^{-1})$$

coincides with the given Q-action. Now G acts on N by conjugation, so that we may form $\Gamma_G^i N$. We may also, for *any* normal subgroup N of a group G, form the lower central G-series, as in Hilton [1974], by the rule

$$(2.6) \qquad \tilde{\Gamma}_G^0 N = N, \ \tilde{\Gamma}_G^{i+1} N = [G, \tilde{\Gamma}_G^i N], \ i \geq 0.$$

We then prove

PROPOSITION 2.5

If N *is a normal subgroup of a group* G, *so that* G *acts on* N *by conjugation, then*

$$\Gamma_G^i N = \tilde{\Gamma}_G^i N.$$

Proof. Again we use induction on i. Then $\tilde{\Gamma}_G^{i+1} N$ is generated by commutators $gag^{-1}a^{-1}$, $g \in G$, $a \in \Gamma_G^i N$ and $gag^{-1}a^{-1} = \overline{g \cdot a}a^{-1} \in \Gamma_G^{i+1} N$. Conversely, $\overline{g \cdot a}ba^{-1}b^{-1} =$ $gag^{-1}a^{-1}aba^{-1}b^{-1} = [g,a][a,b] \in [G, \Gamma_G^i N]$, $g \in G$, $a \in \Gamma_G^i N$, $b \in N$.

PROPOSITION 2.6

Let N *be a* Q-*group and form the split extension (2.4). Then*

$$\Gamma_G^i N = \Gamma_Q^i N.$$

Proof. The argument is again inductive, so we assume $\Gamma_G^i N = \Gamma_Q^i N$. Then, in view of Proposition 2.5, $\Gamma_G^{i+1} N$ is the group generated by the counterimages under μ of the commutators

$$(b,x)(a,1)(b,x)^{-1}(a,1)^{-1}, \ x \in Q, \ a \in \Gamma_Q^i N, \ b \in N$$

$$= (b\overline{xa}b^{-1}a^{-1}, 1).$$

Thus $\Gamma_G^{i+1} N$ is the group generated by the elements $b\overline{xa}b^{-1}a^{-1}$, and this is obviously $\Gamma_Q^{i+1}N$.

In view of Proposition 2.6 we make the following definition.

DEFINITION 2.1

Let N be a Q-group and let $N \rightarrowtail G \twoheadrightarrow Q$ be an extension of Q by N. We say that the extension is *compatible* with the Q-action if[1]

$$\Gamma_G^i N = \Gamma_Q^i N, \quad i \geq 0.$$

Note the compatibility depends only on the *class* of the extension, since, if N is also normal in H and $\phi : G \cong H$ is an isomorphism reducing to the identity on N, then $\Gamma_G^i N = \Gamma_H^i N$.

THEOREM 2.7

Let N *be a* Q*-group and let* $N \rightarrowtail G \twoheadrightarrow Q$ *be an extension of* Q *by* N *compatible with the* Q*-action. Then* G *is nilpotent if and only if* Q *is nilpotent and* N *is* Q*-nilpotent.*

Proof. It is plain that if G is nilpotent of class c, then $\Gamma^c Q = \{1\}$, $\Gamma_G^c N = \{1\}$. Conversely, let $\Gamma^c Q = \{1\}$, $\Gamma_G^d N = \{1\}$. Then $\Gamma^c G \subseteq N$, so that $\Gamma^{c+d} G \subseteq \Gamma_G^d N = \{1\}$. Indeed, we have

$$(2.7) \qquad \max(\text{nil } Q, \text{nil}_Q N) \leq \text{nil } G \leq \text{nil } Q + \text{nil}_Q N.$$

Suppose now that N is nilpotent, and let P be a family of primes. Then we may localize N by $e : N \to N_P$ and if N is a Q-group, there is a unique action of Q on N_P so that e becomes a Q-map. We suppose N_P endowed with this action and prove

---

[1] If N is commutative, then the extension $N \rightarrowtail G \twoheadrightarrow Q$ *determines* an action of Q on N; thus, in this case, we have a finer concept of compatibility.

THEOREM 2.8

$$(\Gamma_Q^i N)_P = \Gamma_Q^i N_P.$$

Proof. It is plain from (2.1) that $e : N \to N_P$ sends $\Gamma_Q^i N$ to $\Gamma_Q^i N_P$, so that e induces

$$e_0 : \Gamma_Q^i N \to \Gamma_Q^i N_P.$$

Thus we must prove that $\Gamma_Q^i N_P$ is P-local. This enables us to embed $(\Gamma_Q^i N)_P$ in $\Gamma_Q^i N_P$ and it will then remain to show that $\Gamma_Q^i N_P \subseteq (\Gamma_Q^i N)_P$.

We prove that $\Gamma_Q^i N_P$ is P-local by induction on i, so we assume that $\Gamma_Q^i N_P$ is P-local for a particular i. Again it suffices to show that $\Gamma_Q^{i+1} N_P$ has p^{th} roots for $p \in P'$. Let $n = p^{f+1}$, where f is as in Blackburn's Theorem (Theorem 1.1). Then, by Theorem 1.3, $[N_P, \Gamma_Q^i N_P]$ has n^{th} roots. If $x \in Q$, $a \in \Gamma_Q^i N_P$, then $a = c^n$, $c \in \Gamma_Q^i N_P$, so

$$\overline{xa}a^{-1} = \overline{xc}^n \cdot c^{-n} = (\overline{xc} \cdot c^{-1})^n u, \; u \in [\Gamma_Q^i N_P, \Gamma_Q^i N_P],$$

whence $\overline{xa}a^{-1} = (\overline{xc} \cdot c^{-1})^n v^n$, $v \in [\Gamma_Q^i N_P, \Gamma_Q^i N_P] \subseteq \Gamma_Q^{i+1} N_P$.

This shows that every element of $\Gamma_Q^{i+1} N_P$ is a product of nth powers, in view of Proposition 2.3, so that $\Gamma_Q^{i+1} N_P$ has p^{th} roots by Blackburn's Theorem.

Now let $e_i \cdot \Gamma_Q^i N \to (\Gamma_Q^i N)_P$ localize $\Gamma_Q^i N$, so that we have the commutative diagram

$$\begin{array}{ccc} & \Gamma_Q^i N & \\ {}^{e_i}\swarrow & & \searrow^{e_0} \\ (\Gamma_Q^i N)_P & \subseteq & \Gamma_Q^i N_P, \end{array}$$

each of e_i, e_0 being obtained by restricting $e : N \to N_P$. We again use induction on i to show that $(\Gamma_Q^i N)_P = \Gamma_Q^i N_P$. For, assuming this true for a particular i, we have, first,

$$[N_P, \Gamma_Q^i N_P] = [N_P, (\Gamma_Q^i N)_P] = [N, \Gamma_Q^i N]_P, \text{ by Theorem 1.2 (iii)}$$

$$\subseteq (\Gamma_Q^{i+1} N)_P.$$

Now if $x \in Q$, $a \in \Gamma_Q^i N_P$, then there exists m in P' with $a^m = ed$, $d \in \Gamma_Q^i N$. Then

$$(\overline{xa}a^{-1})^m = \overline{xa^m}a^{-m}u, \ u \in [\Gamma_Q^i N_P, \Gamma_Q^i N_P] \subseteq (\Gamma_Q^{i+1} N)_P.$$

$$= e(\overline{xd}d^{-1})u \in (\Gamma_Q^{i+1} N)_P.$$

However, $(\Gamma_Q^{i+1} N)_P$ and N_P have unique m^{th} roots and $(\Gamma_Q^{i+1} N)_P \subseteq N_P$. Thus it follows that $\overline{xa}a^{-1} \in (\Gamma_Q^{i+1} N)_P$ whence $\Gamma_Q^{i+1} N_P \subseteq (\Gamma_Q^{i+1} N)_P$, and finally, $\Gamma_Q^{i+1} N_P = (\Gamma_Q^{i+1} N)_P$.

COROLLARY 2.9

If Q *acts nilpotently on* N, *then* Q *acts nilpotently on* N_P; *indeed,*

$$\operatorname{nil}_Q N_P \leqslant \operatorname{nil}_Q N.$$

3. *Nilpotent actions of nilpotent groups*

In this section we suppose that Q is a nilpotent group acting nilpotently on N; it follows from (2.3) that N is itself nilpotent. Moreover, given any extension

(3.1) $$N \rightarrowtail G \overset{\varepsilon}{\twoheadrightarrow} Q$$

compatible with the Q-action, then G is also nilpotent.

THEOREM 3.1

Suppose we may embed (3.1) in

(3.2)
$$\begin{array}{ccccc} N & \rightarrowtail & G & \overset{\varepsilon}{\twoheadrightarrow} & Q \\ \downarrow{\scriptstyle e} & & \downarrow{\scriptstyle \phi} & & \| \\ N_P & \rightarrowtail & H & \overset{\eta}{\twoheadrightarrow} & Q \end{array}$$

Then H *is nilpotent; and the bottom extension in (3.2) is compatible with the action of* Q *on* N_P.

Proof. By a theorem of Philip Hall [1958] (see also Section 10 of Chapter V of Stammbach [1973]) it suffices to show that $H/[N_P,N_P]$ is nilpotent. Now since $e : N \to N_P$ induces $N_{ab} \to N_{Pab}$, which is also P-localizing, it follows that (3.2) induces

(3.3)
$$\begin{array}{ccccc} N_{ab} & \rightarrowtail & G/[N,N] & \twoheadrightarrow & Q \\ \downarrow{\scriptstyle e} & & \downarrow{\scriptstyle \bar{\phi}} & & \| \\ N_{abP} & \rightarrowtail & H/[N_P,N_P] & \twoheadrightarrow & Q \end{array}$$

and, of course, $G/[N,N]$ is nilpotent. Further, the abelianization $\rho : N \twoheadrightarrow N_{ab}$ sends $\Gamma^i_Q N$ onto $\Gamma^i_Q N_{ab}$, and $\Gamma^i_G N$ onto $\Gamma^i_{\bar{G}} N_{ab}$, where $\bar{G} = G/[N,N]$, so that the top extension of (3.3) is compatible with the (nilpotent) action of Q on N_{ab}. It thus follows that we may suppose in (3.2) that N is commutative.

In that case (3.2) determines a (possibly new) Q-action on N by the rule

(3.4) $$x*a = gag^{-1},\ x \in Q,\ a \in N,\ g \in G,\ \varepsilon g = x;$$

and, similarly, a Q-action on N_P by the rule

(3.5) $$x*b = hbh^{-1},\ x \in Q,\ b \in N_P,\ h \in H,\ \eta h = x.$$

Moreover, $e : N \to N_P$ is a Q-map with respect to these actions; and the extensions in (3.2) are *both* compatible with these actions,

$$\overset{*}{\Gamma}{}^i_Q N = \Gamma^i_G N, \quad \overset{*}{\Gamma}{}^i_Q N_P = \Gamma^i_H N_P,$$

where $\overset{*}{\Gamma}{}^i_Q$ is defined with respect to the actions (3.4), (3.5). But then Theorem 2.8 assures us that the Q-action (3.5) is nilpotent; for, since the original Q-action and the new (3.4) Q-action on N are both compatible with (3.1) and the original Q-action is nilpotent, so is the new Q-action. Thus, since Q is nilpotent and operates nilpotently on N_P via (3.5), H is nilpotent.

Reverting to (3.2), it remains to show that

$$\Gamma^i_Q N_P = \Gamma^i_H N_P.$$

Since e plainly maps $\Gamma^i_Q N$ to $\Gamma^i_Q N_P$ and $\Gamma^i_G N$ to $\Gamma^i_H N_P$ and since $\Gamma^i_Q N = \Gamma^i_G N$, it is sufficient, in view of Theorem 2.8, to show that the restriction $e_0 : \Gamma^i_G N \to \Gamma^i_H N_P$, of e, P-localizes. Now it follows from (3.2) and Theorem 6.5 of Hilton [1973] that $\phi : G \to H$ is P-bijective. Thus, arguing by induction on i, we obtain, by a slight modification of the argument of Proposition 1.6 of Hilton [1974], that e_0 is P-bijective, and then Corollary 1.4 ensures that $\Gamma^i_H N_P$ is P-local. This completes the proof of the theorem.

We now show that such a diagram (3.2) can be obtained from (3.1) and is essentially unique. Our procedure is to localize (3.1) and then use $e : Q \to Q_P$ to pull-back the localization of (3.1). The full diagram is as follows.

(3.6)

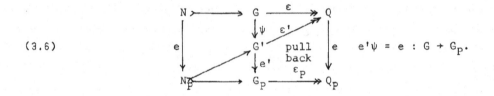

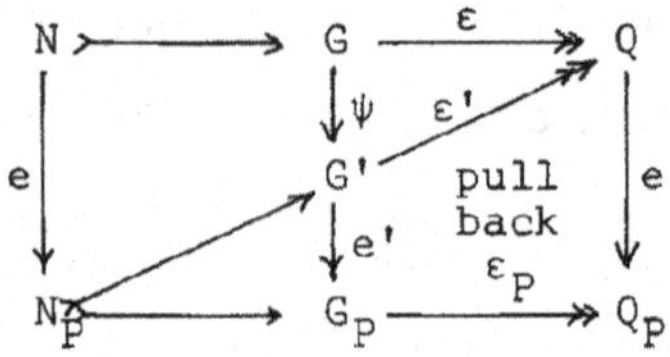

$e'\psi = e : G \to G_P.$

Now suppose that (3.2) is given. Then H is nilpotent and $\phi : G \to H$ is P-bijective. It follows that the localization of H is just G_P, that is, we have a commutative diagram

(3.7)
$$\begin{array}{ccccc} N_P & \rightarrowtail & H & \overset{\eta}{\text{———}} & Q \\ \| & & \Big| e'' & & \Big| e \\ N_P & \rightarrowtail & G_P & \overset{\varepsilon_P}{\text{———}} & Q_P \end{array}$$

Thus, in view of the pull-back in (3.6) we have a commutative diagram

$$\begin{array}{ccccc} N_P & \rightarrowtail & H & \overset{\eta}{\twoheadrightarrow} & Q \\ \| & & \downarrow \theta & & \| \\ N_P & \rightarrowtail & G' & \overset{\varepsilon'}{\twoheadrightarrow} & Q \end{array},$$

establishing the existence and essential uniqueness of the diagram (3.2) arising out of (3.1).

Now suppose in particular that (3.1) is the split extension of Q by N with splitting $\sigma : Q \to G$. Then in (3.6) we have a splitting $\sigma_P : Q_P \to G_P$ which determines an action of Q_P on N_P. We call this the Q_P-action on N_P *induced* by the given Q-action on N; note that it is only defined for nilpotent actions of nilpotent groups. Then (3.6) also yields a splitting $\sigma' : Q \to G'$, with $\varepsilon'\sigma' = 1$, obtained by lifting σ_P; however, the Q-action on N_P determined by σ' is just the action obtained by localizing the action of Q on N, as the reader may easily check.

THEOREM 3.2

Let the nilpotent group Q *act nilpotently on* N, *inducing actions of* Q *on* N_P *and of* Q_P *on* N_P. *Then these induced actions are also nilpotent. Indeed, referring to diagram (3.6), based on the extension (3.1) compatible with the* Q-*action on* N,

$$(3.8) \qquad (\Gamma_Q^i N)_P = \Gamma_Q^i N_P = \Gamma_{G'}^i N_P = \Gamma_{Q_P}^i N_P = \Gamma_{G_P}^i N_P .$$

Proof. We have already established the first two equalities in (3.8), the second in the course of the proof of Theorem 3.1. Thus it only remains to establish the last two equalities since the nilpotency of G_P will then immediately imply the nilpotency of the (induced) action of Q_P on N_P.

Now in (3.6) $e' : G' \to G_P$ P-localizes; this immediately establishes $\Gamma_{G'}^i N_P = \Gamma_{G_P}^i N_P$ in view of Theorem 1.2 (iii). The final equality we prove is $(\Gamma_Q^i N)_P = \Gamma_{Q_P}^i N_P$. However, to prove this inequality, we again suppose that (3.1) is split by $\sigma : Q \to G$. Then, as remarked, the bottom extension of (3.6) is split by σ_P, and, by Proposition 2.6, we have, in this case,

$$\Gamma_G^i N = \Gamma_Q^i N, \; \Gamma_{G_P}^i N_P = \Gamma_{Q_P}^i N_P .$$

Since we know that $(\Gamma_Q^i N)_P = \Gamma_G^i N_P$ (for *any* extension compatible with the Q-action) it follows that $(\Gamma_Q^i N)_P = \Gamma_{Q_P}^i N_P$ and the theorem is proved.

Let Q be nilpotent and let $A(Q,N)$ be the set of nilpotent actions of Q on the nilpotent group N. Each such nilpotent action induces a nilpotent action of Q on N_P and of Q_P on N_P. We thus have functions

(3.9)
$$\begin{array}{ccc} A(Q,N) & \xrightarrow{\nu} & \\ \downarrow \mu & & A(Q_P,N_P) \\ A(Q,N_P) & \xrightarrow{\lambda} & \end{array}$$

and the diagram (3.9) commutes; this is essentially the remark preceding the statement of Theorem 3.2. For $e : Q \to Q_P$ plainly induces

$$e^* : A(Q_P,N_P) \to A(Q,N_P);$$

our remark amounts to $e^*\nu = \mu$; and we now claim

THEOREM 3.3

$\lambda : A(Q,N_P) \to A(Q_P,N_P)$ *is bijective with inverse* e^*.

Proof. The proof is just as in Hilton [1974; Theorem 2.5]. We just take the bottom part of (3.6), in the case of a split extension,

$$\begin{array}{ccccc} N_P & \rightarrowtail & G' & \twoheadrightarrow & Q \\ \| & & \downarrow e' & & \downarrow e \\ N_P & \rightarrowtail & G_P & \twoheadrightarrow & Q_P, \end{array}$$

and observe that we may either regard the top extension as obtained from the bottom by pulling back by $e : Q \to Q_P$, or the bottom extension as obtained from the top by P-localizing. Thus we pair off nilpotent actions of Q on N_P and nilpotent actions of Q_P on N_P.

4. *Fixpoint sets for nilpotent actions*

We again suppose we have a nilpotent group Q acting nilpotently on a group N, and we define N^Q to be the subgroup of N

consisting of those elements $a \in N$ such that

$$x \cdot a = a, \text{ for all } x \in Q.$$

We prove the following theorem relating to the fixpoint subgroups for the Q-action on N and the induced Q-action on N_P and Q_P-action on N_P. (Note that in part (i) of the theorem neither Q nor the action is required to be nilpotent.)

THEOREM 4.1

(i) For any Q*-action on the nilpotent group* N, N_P^Q *is* P*-local and* $(N^Q)_P \subseteq N_P^Q$. *If* Q *is finitely generated then* $(N^Q)_P = N_P^Q$.

(ii) For any nilpotent action of the nilpotent group Q *on* N_P, *we have* $N_P^{Q_P} = N_P^Q$.

Proof. (i) The proof of the first assertion is just as for Theorem 2.8 (i) of Hilton [1974]. To prove the second assertion we remark that it suffices to show that $e_0 : N^Q \to N_P^Q$, obtained by restricting $e : N \to N_P$, is P-surjective. Thus let $a \in N_P^Q$. Then there exists $n \in P'$ such that $a^n = eb$ for some $b \in N$. Let $Q = gp\,(x_1, x_2, \ldots, x_k)$. Then

$$e(x_i b) = x_i a^n = a^n = eb,$$

so that $x_i b = bu_i$, where $u_i^{m_i} = 1$, for some $m_i \in P'$, $i = 1, 2, \ldots, k$. Let nil $N \leq c$. Then, by Corollary 6.2 of Hilton [1973], $x_i b^{m_i^c} = b^{m_i^c}$, so that, if $m = (m_1 m_2 \ldots m_k)^c$, then $m \in P'$ and

$$x_i b^m = b^m, \; i = 1, 2, \ldots, k.$$

This shows that $b^m \in N^Q$ and $eb^m = a^{mn}$, $mn \in P'$, showing that e_0 is

P-surjective.

(ii) It is trivial that $N_P^{Q_P} \subseteq N_P^Q$ (recall that, by Theorem 3.3, $e : Q \to Q_P$ induces a bijection between nilpotent actions of Q_P on N_P and nilpotent actions of Q on N_P). Thus it remains to show that $N_P^Q \subseteq N_P^{Q_P}$. Now, given an action of any group Q on any group N, we may form the split extension

$$N \rightarrowtail G \twoheadrightarrow Q \qquad \overset{\sigma}{\longleftarrow}$$

and it follows immediately from (2.5) that, regarding N as a subgroup of G, $N^Q = N \cap Z(\sigma Q)$. We will also think of Q as embedded in G by σ, so that we may write

(4.1) $$N^Q = N \cap Z(Q).$$

In fact, we will take a given nilpotent action of the nilpotent group Q on N_P, form the split extension, and P-localize; we obtain

(4.2) $$\begin{array}{ccccc} N_P & \rightarrowtail & G & \rightleftarrows & Q \\ \| & & \downarrow e & & \downarrow e \\ N_P & \rightarrowtail & G_P & \rightleftarrows & Q_P \end{array}$$

Then $N_P^Q = N_P \cap Z(Q)$, $N_P^{Q_P} = N_P \cap Z(Q_P)$. Under the localization $e : G \to G_P$, we have that $Z(Q)$ is mapped to $Z(Q_P)$ (Theorem 1.5). Since e reduces to the identity on N_P it follows that $N_P^Q \subseteq N_P^{Q_P}$, and the theorem is proved.

Remark. The second part of Theorem 4.1 (i) provides a considerable strengthening of Corollary 2.9 of Hilton [1974], even when N is commutative. In particular we no longer require Q nilpotent; but we also no longer require nilpotent action, nor that N be finitely generated.

5. *Appendix*

In this appendix we first give a new proof of Theorem 1.2 (iii) which does not involve Theorem 1.3. Let N_c be the category of nilpotent groups of class $\leq c$. Then if $H, K \in N_c$ their coproduct in N_c is

$$H \sqcup K = H * K/\Gamma^c(H*K),$$

where $H * K$ is the free product of H and K. We proved in Hilton [1974] that $e \sqcup e : H \sqcup K \to H_P \sqcup K_P$ is P-bijective. Now H, K embed naturally in $H \sqcup K$ (as retracts). Let $\{H,K\}$ be the commutator group of H and K as subgroups of $H * K$ and let (H,K) be the commutator group of H and K as subgroups of $H \sqcup K$. Since $\{H,K\}$ is the kernel of the natural map $H * K \to H \times K$, it follows that (H,K) is the kernel of the natural map $H \sqcup K \to H \times K$,

$$(5.1) \qquad (H,K) \rightarrowtail H \sqcup K \twoheadrightarrow H \times K.$$

PROPOSITION 5.1

(H_P,K_P) *has* p^{th} *roots*, $p \in P'$.

Proof. We apply (5.1) with H_P, K_P replacing H, K; thus,

$$(H_P,K_P) \rightarrowtail H_P \sqcup K_P \twoheadrightarrow H_P \times K_P.$$

An easy application of Blackburn's Theorem shows that $H_P \sqcup K_P$ has p^{th} roots for $p \in P'$. Since $H_P \times K_P$ is P-local, it follows immediately that (H_P,K_P) has p^{th} roots.

COROLLARY 5.2

Let G *be nilpotent,* H, K *subgroups of* G. *Then* $[H,K]_P = [H_P,K_P]$.

Proof. Assume nil $G \leq c$. Then we know that $(e,e) : (H,K) \to (H_P,K_P)$ is P-bijective and the natural map $H \sqcup K \to G$ maps (H,K) onto $[H,K]$. We thus have the diagram

$$\begin{array}{ccc} (H,K) & \twoheadrightarrow & [H,K] \\ \downarrow{\scriptstyle (e,e)} & & \downarrow{\scriptstyle e_0} \\ (H_P,K_P) & \twoheadrightarrow & [H_P,K_P] \end{array}$$

where e_0 is the restriction of $e : G \to G_P$. Then e_0 is certainly P-injective. Since (e,e) is P-surjective, so is e_0; and since (H_P,K_P) has p^{th} roots, $p \in P'$, so has $[H_P,K_P]$. It follows that $[H_P,K_P]$ is P-local (since it is a subgroup of G_P), so that e_0 is the P-localizing map.

Remark. To prove that $H_P \sqcup K_P$ is the P-localization of $H \sqcup K$ we would need to know that it is P'-torsionfree. Our argument shows that this is equivalent to (H_P,K_P) being P'-torsionfree. At present all we can say is that $(H \sqcup K)_P$ is obtained from $H_P \sqcup K_P$ by factoring out the P'-torsion, and similarly $(H,K)_P$ is obtained from (H_P,K_P) by factoring out the P'-torsion. Of course H_P, K_P are embedded naturally in $(H_P \sqcup K_P)_P = (H \sqcup K)_P$ as retracts and $(H,K)_P$ is the commutator group of H_P and K_P as subgroups of $(H \sqcup K)_P$.

Consider now Corollary 1.4. Though this was adequate for the application we had in mind (Theorem 3.1, 3.1), we may strengthen the corollary considerably by no longer requiring that G itself be nilpotent.

THEOREM 5.3

Let K *be a normal* P-*local nilpotent subgroup of the group* G *and let* H *be a subgroup of* G. *Then* [H,K] *is* P-*local.*

Proof. Since $[H,K] \subseteq K$, it is sufficient to show that [H,K] has p^{th} roots, $p \in P'$. Now consider [[H,K],K]. We may apply Theorem 1.3 (with the roles of G, H in that theorem being played by K, [H,K] respectively) to infer that [[H,K],K] has p^{th} roots. Reverting to the proof of Theorem 1.3, set nil [H,K] = c, $f = f(p,c)$ (in Theorem 1.1), $n = p^{f+1}$. Then, for $a \in H$, $k \in K$, we set $k = b^n$, $b \in K$, and

$$[a,k] = [a,b^n] = [a,b]^n u, \quad u \in [[H,K],K]$$
$$= [a,b]^n v^n, \quad v \in [[H,K],K] \subseteq [H,K].$$

Thus we infer, as in the proof of Theorem 1.3, that every element of [H,K] is a product of nth powers, and is therefore itself a pth power.

Remark. Of course, had we merely assumed K normal nilpotent with p^{th} roots, we would have inferred that [H,K] has p^{th} roots. However, to obtain this inference, it is sufficient to assume that K has p^{th} roots and is contained in a normal nilpotent subgroup of G.

BIBLIOGRAPHY

BAUMSLAG, G.

[71] *Lecture Notes on Nilpotent Groups*, A.M.S. Regional Conference Series No. 2, 1971.

BLACKBURN, N.

[65] Conjugacy in nilpotent groups, Proc. Amer. Math. Soc. 16 (1965), 143-148.

HALL, P.

[58] Some sufficient conditions for a group to be nilpotent, Ill. J. Math. 2 (1958), 787-801.

HILTON, P.

[73] Localization and cohomology of nilpotent groups, Math. Zeits. 132 (1973), 263-286.

[74] Remarks on the localization of nilpotent groups, Comm. Pure and App. Math. (1974).

HILTON, P., G. MISLIN and J. ROITBERG

[73] Homotopical localization, Proc. London Math. Soc. 26 (1973), 693-706.

STAMMBACH, U.

[73] *Homology in Group Theory*, Lecture Notes in Mathematics 359, Springer (1973).

SULLIVAN, D.

[70] Geometric topology, part 1: Localization, periodicity and Galois symmetry, MIT, June 1970. (mimeographed notes)

* * *

Battelle Seattle Research Center, Seattle, Washington, USA

Case-Western Reserve University, Cleveland, Ohio, USA

Department of Mathematics, Monash University, Melbourne, Australia

STRUCTURE THEOREMS FOR INVERSE SEMIGROUPS

R. McFadden

Just as the concept of a group is a characterization of a semigroup of permutations on a set, with the inverse of each of the permutations also being in the semigroup, the concept of an inverse semigroup characterizes certain semigroups of one-to-one mappings and their inverses, as follows. For a given set X, instead of considering only permutations of X, consider semigroups of one-to-one mappings from one subset of X onto another. If α has domain A and β has domain B, then $\alpha\beta$ has domain $A \cap B\alpha^{-1}$, and $x(\alpha\beta)$ is just $(x\alpha)\beta$. The concept of an inverse semigroup is a characterization of such semigroups which with $\alpha : A \to B$ also contain $\alpha^{-1} : B \to A$. The set $\mathcal{I}_X$ of *all* one-to-one mappings of subsets of X onto subsets of X is such a semigroup. It is called the symmetric inverse semigroup on X, and is the basic model for all inverse semigroups. Every inverse semigroup S is isomorphic to a subsemigroup of $\mathcal{I}_X$ for some X; in fact, one may take X = S (Clifford and Preston [67]). This provides an excellent model for inverse semigroups.

Inverse semigroups may be considered as a tool for investigating local properties of X, as opposed to global properties. For example, if F is a field and K a normal extension of F, as well as studying the Galois group of K over F, one might consider the inverse semigroup of F-isomorphisms between subfields of K containing F. Or if $\mathcal{A}$ is a category in which for each object A, any two subobjects of A have an intersection, then for each $A \in \mathcal{A}$ there is an inverse semigroup $I(A)$ consisting of isomorphisms between subobjects of A. One may ask: for what categories does $I(A) \approx I(B)$ imply $A \simeq B$? The implication does hold for finite abelian groups (Preston [73]), but not for abelian groups; there exist non-isomorphic abelian groups A, B of rank 1, such that $I(A) \approx I(B)$.

Axiomatic treatments of inverse semigroups were first provided by Vagner and Preston [67]. The abstract definition is that an inverse semigroup is a semigroup S in which

$$\text{for each } a \in S \quad \exists!\, x \in S \quad \text{such that } axa = a,\ xax = x.$$

This unique element is denoted by a^{-1} and called the inverse of a; we have $(a^{-1})^{-1} = a$ and $(ab)^{-1} = b^{-1}a^{-1}$. Inverse semigroups form a variety and knowledge of the variety is growing. The free object is now known (Scheiblich [to appear]), and the variety has amalgamation, even strong amalgamation (Hall [73]).

Although aa^{-1} and $a^{-1}a$ need not be equal, each is an idempotent. The idempotents in an inverse semigroup have an important property: they commute. This makes it possible to consider the set $E(S)$, or simply E, of idempotents of an inverse semigroup S, as a semilattice, in which $e \wedge f = ef$. Clearly, for each $e \in E$ there is a subgroup H_e of S consisting of those elements of eSe which are invertible with respect to e. The H_e are called the maximal subgroups of S; they are disjoint.

An important part of the theory of inverse semigroups is to describe the structure of a given inverse semigroup in terms of groups and/or semilattices (themselves, of course, inverse semigroups).

There is a wide literature on this problem; I shall describe two approaches to it.

To begin, consider an example.

Let E be a given semilattice, and let
$T_E = \{\alpha \in \mathcal{I}_E : \alpha$ is an isomorphism between principal ideals of $E\}$.

Then T_E is an inverse subsemigroup of $\mathcal{J}_E$. Its idempotents are the identity mappings on principal ideals of E, and so the semilattice of idempotents of T_E is in fact isomorphic to E. This is a very useful construction for generating examples of inverse semigroups.

For an arbitrary inverse semigroup, define the relations $\mathcal{R}$, $\mathcal{L}$, $\mathcal{H}$, $\mathcal{D}$ and $\mathcal{J}$ (Green's relations) by:

$a\mathcal{R}b$ if and only if $aS = bS$, $a\mathcal{L}b$ if and only if $Sa = Sb$, $\mathcal{H} = \mathcal{R} \cap \mathcal{L}$, $\mathcal{D} = \mathcal{R} \vee \mathcal{L}$, $a\mathcal{J}b$ if and only if $SaS = SbS$;

each is an equivalence relation on S. In fact, $\mathcal{R}$ and $\mathcal{L}$ commute, and so $\mathcal{D} = \mathcal{R} \circ \mathcal{L}$. Within a given $\mathcal{D}$-class the $\mathcal{H}$-classes containing idempotents are isomorphic groups; they are the maximal subgroups of S. Green's equivalences may be defined for arbitrary semigroups, and these results still hold. For inverse semigroups, they take on a particularly nice form:

$a\mathcal{R}b$ if and only if $aa^{-1} = bb^{-1}$; $a\mathcal{L}b$ if and only if $a^{-1}a = b^{-1}b$, $a\mathcal{D}b$ if and only if $\exists c \in S$, $aa^{-1} = cc^{-1}$, $c^{-1}c = b^{-1}b$.

Within a given $\mathcal{D}$-class the $\mathcal{R}$- and $\mathcal{L}$-classes are in (1-1)-correspondence, and each contains a unique idempotent.

For each $e \in E$ and each $a \in S$, $a^{-1}ea \in E$, and in fact $\theta_a : E \to E$ defined by $e\theta_a = a^{-1}ea$ is an endomorphism of E. In general it is not an isomorphism, but if θ_a is restricted to Eaa^{-1}, the principal ideal of E generated by aa^{-1}, it is an isomorphism of Eaa^{-1} onto $Ea^{-1}a$. Therefore $\theta : a \mapsto \theta_a$ maps S into T_E, and in fact θ is a homomorphism of S into T_E (Munn [66]).

By definition, $a\theta = b\theta$ if and only if

$$aa^{-1} = bb^{-1},\ a^{-1}a = b^{-1}b \text{ and } a^{-1}ea = b^{-1}eb$$
$$\forall\, e \leqslant aa^{-1}.$$

It follows that the congruence

$$\mu = \{(a,b) \in S \times S : a^{-1}ea = b^{-1}eb\ \forall\, e \in E\}$$

determined by θ is contained in $\mathcal{H}$, and in fact μ is the maximum idempotent separating congruence on S.

Every homomorphic image of an inverse semigroup is also an inverse semigroup, and we have that

$$E \approx E(S/\mu) \approx E(T_E);$$

this last is expressed by saying that S/μ is a *full* subsemigroup of T_E. Further, if we say that an inverse semigroup is *fundamental* if $\mu = \iota$, we have half of a theorem due to Munn [70].

THEOREM 1

An inverse semigroup with semilattice E *is fundamental if and only if it is isomorphic to a full inverse subsemigroup of* T_E.

In particular, T_E is fundamental.

Since the canonical homomorphism $S \to S/\mu$ is idempotent separating, its kernel is a union of groups. The problem of constructing idempotent separating extensions of unions of groups by inverse semigroups has been solved (Coudron [68], D'Alarcao [69]), so one can construct all inverse semigroups if one knows all fundamental inverse semigroups. There is a cohomology theory for such extensions, at least in the case where the kernel is a union of abelian groups (Lausch [73]).

An interesting category is the following. For a given semilattice E, the objects are inverse semigroups with semilattice of idempotents E; the morphisms are homomorphisms which preserve E.

Munn also characterized, in terms of E and certain subsemigroups of T_E, those inverse semigroups which consist of a single $\mathcal{D}$-class, or of a single $\mathcal{J}$-class. Recall that if S is $\mathcal{D}$-simple, then $\forall a, b \in S \; \exists c \in S$, with $aa^{-1} = cc^{-1}$, $c^{-1}c = b^{-1}b$. Interpreted in terms of mappings, this is equivalent to $\theta_c : Eaa^{-1} \to Ec^{-1}c$ and $\theta_b : Ebb^{-1} \to Ec^{-1}c$ being isomorphisms; in other words, $\theta_c\theta_b^{-1}$ is an isomorphism of Eaa^{-1} onto Ebb^{-1}. Conversely, if every two principal ideals of E are isomorphic then T_E is $\mathcal{D}$-simple (and has semilattice E).

Let E be a given semilattice. Call an inverse subsemigroup S of T_E *transitive* if for each $e, f \in E \; \exists \alpha \in S$ such that α has domain Ee and codomain Ef. Further, E is *uniform* if and only if T_E is transitive. Then (Munn [70]) :

THEOREM 2

(i) A semilattice E *is uniform if and only if it is (to within isomorphism) the semilattice of a* $\mathcal{D}$*-simple inverse semigroup.*

(ii) An inverse semigroup S *with semilattice* E *is fundamental and* $\mathcal{D}$*-simple if and only if it is isomorphic to a transitive inverse subsemigroup of* T_E.

There is an analogous theorem for *subuniform* semilattices (T_E is *subtransitive*, in the sense that $\forall\, e, f \in E\ \exists\, \alpha \in T_E$ with domain $\alpha = Ee$, codomain $\alpha \subseteq Ef$) and $\mathcal{J}$-simple inverse semigroups. The existence of a zero in S complicates the problem, but Munn has successfully treated that case too.

Just as μ is the largest congruence on an inverse semigroup S which *separates* idempotents, so there is a smallest congruence σ which *identifies* all the idempotents of S. It is defined by:

$$(a,b) \in \sigma \text{ if and only if } \exists e \in E,\ ea = eb,$$

and since $aa^{-1} \in E\ \forall\, a \in S$, it follows that S/σ is the maximum group homomorphic image of S.

Consider now a partially ordered set X, a subsemilattice Y of X which is also an (order)ideal of X, and a group G which acts on X by order automorphism. On the set

$$P(G,X,Y) = P = \{(a,g) \in Y \times G : g^{-1}a \in Y\}$$

define

$$(a,g)(b,h) = (a \wedge gb,\ gh).$$

Then P is an inverse semigroup, with $(a,g)^{-1} = (g^{-1}a, g^{-1})$, and $E(P) = \{(a,1) : a \in Y\} \approx Y$. We call it a P-semigroup. These semigroups are explicitly given. In particular, it is easy to describe Green's relations on them. Thus, for example, there is a one-to-one correspondence between the set of orbits of G and the set of $\mathcal{D}$-classes of $P(G,X,Y)$.

In general, the semilattice of idempotents of the symmetric inverse semigroup $\mathcal{I}_X$ on a set X is isomorphic to 2^X, the set of subsets of X, so there is a natural action of the symmetric group S_X on the idempotents of $\mathcal{I}_X$, defined by:

$$\alpha \cdot A = A\alpha^{-1} \quad \text{for each} \quad \alpha \in S_X,\ A \in 2^X.$$

Thus $P(S_X,2^X,2^X)$ consists of all pairs $(A,\alpha) \in 2^X \times S_X$ under

$$(A,\alpha)(B,\beta) = (A \cap \alpha \cdot B,\ \alpha\beta) = (A \cap B\alpha^{-1},\ \alpha\beta).$$

This is strongly reminiscent of multiplication in $\mathcal{I}_X$. In fact, if we define $\psi : P \to \mathcal{I}_X$ by $(A,\alpha)\psi = \alpha|_A$, then ψ is a homomorphism, and is *idempotent separating*. If X is finite ψ is onto, and in this case $\mathcal{I}_X$ is an idempotent separating homomorphic image of a P-semigroup.

If X is infinite, this approach may be amended by adding to X another, disjoint, set of the same cardinality, and cutting down on the size of Y; this leads to (McAlister [to appear][a]):

THEOREM 3

Every inverse semigroup is an idempotent separating homomorphic image of a full inverse subsemigroup of a P*-semigroup.*

At this point the duality between Munn's and McAlister's approaches is very strong, right down to the questions: what are the full inverse subsemigroups of T_E ? and what are the full inverse subsemigroups of P(G,X,Y)?

In a remarkable paper, soon to appear, McAlister [to appear][b] has answered the second question: they are all P-semigroups! Thus every inverse semigroup is an idempotent separating homomorphic image of a P-semigroup, and since these homomorphisms are known (Clifford and Preston [61,67], in theory at least the structure of arbitrary inverse semigroups is known.

The key to McAlister's result is that in a P-semigroup E is a *complete* σ-class; one says that P is *proper*. The class of proper inverse semigroups is a wide one; for example, it contains the free inverse semigroup on a set (McAlister and McFadden [to appear]). But not all inverse semigroups are proper ($\mathcal{I}_X$ is not), and homomorphic images of proper inverse semigroups need not be proper.

The property of propriety is hereditary, so by Theorem 3 every inverse semigroup is an idempotent separating homomorphic image of a proper inverse semigroup, and it remains to prove (McAlister [to appear][b]):

THEOREM 4

Every proper inverse semigroup is a P-semigroup.

This is a difficult theorem to prove, and the proof is also intricate. However, I shall try to indicate the main ingredients.

Consider a semigroup P(G,X,Y) = P. Not all the elements of $Y \times G$ appear in P : (a,g) appears if and only if $g^{-1}a \in Y$. This has consequences for G and for X. The only elements g of G which appear in elements of P are those for which $gY \cap Y \neq \Box$. The only elements of X which appear in elements of P are those of the form gb, some $b \in Y$, $g \in G$. So if we let

$$G' = \{g \in G : gY \cap Y \neq \square\}$$

and $X' = G \cdot Y$, then G' is a group acting on the partially ordered set X' by order automorphisms, and Y is a subsemilattice ideal of X'. Further, $P(G,X,Y) = P(G',X',Y)$, so in general we may require that $G \cdot Y = X$ and $gY \cap Y \neq \square$ for each $g \in G$.

Now let $((a,g),(b,h)) \in \sigma$. By definition, $\exists (c,1) \in P$ such that $(c,1)(a,g) = (c,1)(b,h)$, and this is true if and only if $g = h$. Hence

$$P/\sigma \approx \{g \in G : (a,g) \in P, \text{ some } a \in Y\} = G;$$

that is, G is isomorphic to the maximum group homomorphic image of P.

Suppose now that S is a proper inverse semigroup with semilattice of idempotents E and maximum group homomorphic image $G = S/\sigma$. If there is a semigroup $P(H,X,Y) \approx S$, we know that $Y \approx E$ and $H \approx G$; the main problem is to find X.

Let D_i, $i \in I$, be the $\mathcal{D}$-classes of S, and pick an idempotent $f_i \in D_i$ for each $i \in I$; denote by H_i the $\mathcal{H}$-class of f_i. For each $i \in I$ pick representatives r_{iu} of the $\mathcal{H}$-classes in the $\mathcal{R}$-class of f_i, with f_i representing H_i; denote this set of representatives by E_i. Then:

(i) each element of S can be uniquely expressed in the form $r_{iu}^{-1} h_i r_{iv}$, where $r_{iu}, r_{iv} \in E_i$, $h_i \in H_i$ for some $i \in I$;

(ii) $E(S) = \{r_{iu}^{-1} r_{iu} : i \in I\}$; they are all distinct.

This type of co-ordinatization is valid for any inverse semigroup. For proper inverse semigroups it is particularly useful because these are exactly the inverse semigroups for which the *canonical homomorphism* $\sigma^{\natural} : S \to S/\sigma$ is *one-to-one on* $\mathcal{R}$*-classes.* They are therefore those *which can be co-ordinatized by* $a \leftrightarrow (aa^{-1}, a\sigma^{\natural})$.

Let $G_i = H_i\sigma^{\natural}$; since S is proper, $G_i \approx H_i$. Write $g_i = h_i\sigma^{\natural}$, $k_{iu} = r_{iu}\sigma^{\natural}$; the k_{iu} are distinct for each $i \in I$.

Let $B_{ij} = \{k_{ju} : r_{ju}^{-1} r_{ju} \leq f_i\}$. Given B_{ij} we can recover Ef_i since there is a unique element in the $\mathcal{R}$-class of f_j mapped onto a given $k \in B_{ij}$.

First, one shows that:

$$r_{iu}^{-1} r_{iu} \geq r_{jv}^{-1} r_{jv} \Leftrightarrow \quad k_{jw} \in B_{ij},\ k_{jv} \equiv k_{jw}\, k_{iu} \pmod{G_j},$$

and a candidate for Y appears.

Take $Y = \{(i, G_i k_{iu}) : i \in I,\ k_{iu} \in E_i\sigma\}$ under

$$(i, G_i x) \geq (j, G_j y) \Leftrightarrow \quad k_{jw} \in B_{ij},\ y \equiv k_{jw} x \pmod{G_j},$$

$$X = \cup\{\{i\} \times G/G_i : i \in I\}$$

with the same ordering and define

$$g \cdot (i, G_i x) = (i, G_i x\, g^{-1}) \quad \text{for} \quad g \in G.$$

Then X is a partially ordered set, Y is a subsemilattice ideal of X, G acts on X by order automorphisms and, recalling that S is co-ordinatized by $a \leftrightarrow (aa^{-1}, a\sigma^{\natural})$, it follows that $\psi : S \to P(G,X,Y)$ defined by

$$(r_{iu}^{-1}\, h_i\, r_{jv})\psi = ((i, G_i\, k_{iu}),\ k_{iu}^{-1}\, g_i\, k_{iv})$$

is an isomorphism of S onto P.

Further, P is unique up to isomorphism of G, order isomorphism of X, and equivalence of group actions.

BIBLIOGRAPHY

CLIFFORD, A.H. and G.B. PRESTON

[61,67] Algebraic theory of semigroups, Math. Surveys 7, Vols.1 and 2, Providence, R.I., 1961, 1967.

COUDRON, A.

[68] Sur les extensions des demigroupes reciproques, Bull. Soc. Roy. Sci., Liège 37, 1968 , 409-419.

D'ALARCAO, H.

[69] Idempotent separating extensions of inverse semigroups, J. Aust. Math. Soc. 9, 1969, 211-217.

HALL, T.E.

[73] Inverse semigroups and the amalgamation property, Monash University Algebra Paper 1, 1973.

LAUSCH, H.

[73] Cohomology of inverse semigroups, Monash University Algebra Paper 6, 1973.

McALISTER, D.B.

[to appear][a] Groups, semilattices and inverse semigroups, Trans. Amer. Math. Soc., to appear.

McALISTER, D.B.

[to appear][b] Groups, semilattices and inverse semigroups II, Trans. Amer. Math. Soc.,to appear.

McALISTER, D.B. and R. McFADDEN

[to appear] Zig-zag representations and inverse semigroups, Journal of Algebra, to appear.

MUNN, W.D.

[66] Uniform semilattices and bisimple inverse semigroups, Q. J. Math. Oxford (2) 17, 1966, 151-159.

[70] Fundamental inverse semigroups, Q.J. Math. Oxford (2), 1970, 157-170.

PRESTON, G.B.

[73] Inverse semigroups; some open questions, Proceedings of a symposium on inverse semigroups and their generalizations, Northern Illinois University, 1973.

SCHEIBLICH, H.E.

[to appear] Free inverse semigroups, to appear.

* * *

Department of Mathematics, Monash University, Clayton , Australia

RECURSION THEORY AND ALGEBRA

G. Metakides and A. Nerode*

1.

Algebra is replete with "constructions". Some are constructions in the sense of recursion theory, some are not. If one is not, then we would like to determine whether that particular "construction" can be replaced by a recursive construction which achieves the same purpose, or whether such a replacement is in principle impossible. The work in progress reported on here has a semi-constructionist motivation. This motivation is to develop machinery for answering these questions in natural cases. Out of this arises a new and non-trivial development of theories of recursively enumerable algebraic structures. It is analogous to but not a corollary of the contemporary theory of recursively enumerable sets. The tools employed are the working tools of the contemporary recursion theorist. These tools apply as powerfully here as in the more traditional context. We may penetrate recursively enumerable algebraic structures to any desired depth – for instance the game theoretic approach to the priority method (Lachlan [70], Yates [74]) mixes well with the requirements imposed by algebra.

---

* Manuscript dated July 1, 1974. This is an expanded version of the talk delivered by A. Nerode to a general mathematical audience at the January, 1974 Recursive Model Theory Symposium at Monash University (Melbourne, Australia).

2. *An example*

We begin with a construction from elementary linear algebra. Suppose I is an independent subset of a vector space V. How is an independent set $v_{n_0}, v_{n_1}, \ldots$ constructed which is a basis of V mod (I)? List V as $v_0, v_1, \ldots$. Search for the first $v_n \not\equiv 0$ mod (I), call it v_{n_0}. Now search for the first $v_n \not\equiv 0$ mod $(I \cup \{v_{n_0}\})$, call it v_{n_1}. Then $v_{n_0}, v_{n_1}, \ldots$ is the required basis mod (I). To set a recursion theoretic stage we assume that V is a recursively presented infinite dimensional vector space with recursive base over a recursive field and that $v_0, v_1, \ldots$ recursively enumerates V. There is an obvious sufficient condition which will insure that $v_{n_0}, v_{n_1}, \ldots$ is a recursively enumerable set. Call a subspace W of V *decidable* if there is a recursive procedure which, when applied to any finite sequence $v_{i_0} \ldots v_{i_n}$ of vectors in V, decides in a finite number of steps whether or not $v_{i_0} \ldots v_{i_n}$ is an independent sequence of vectors mod W. If the subspace W generated by I is decidable, then $v_{n_0}, v_{n_1}, \ldots$ is recursively enumerable.

How much can the assumption that the subspace W be decidable be weakened and still entail the conclusion that there is a recursively enumerable basis $v_{n_0}, v_{n_1}, \ldots$ mod W? For example, is it sufficient to assume that the subspace W is a recursive *set* ? Answer: no.

With the aid of the finite injury priority method we produce a recursively enumerable independent set I with the following properties: (i) The subspace W generated by I is a recursive set;

(ii) V mod W is infinite dimensional;

(iii) Whenever J is a recursively enumerable independent

set and $J \supseteq I$, then $J - I$ is finite.

There can be no recursively enumerable set $v_{n_0}, v_{n_1}, \ldots$ which constitutes a basis mod W. For then $J = I \cup \{v_{n_0}, v_{n_1}, \ldots\}$ satisfies (iii), so $J - I$ is finite, so $v_{n_0}, v_{n_1}, \ldots$ is a finite sequence, so V mod W is finite dimensional, contrary to (ii). This I is a recursively enumerable set which cannot be extended to a recursively enumerable basis.

3. *Lattices of recursively enumerable structures*

With every recursively presented model $\mathfrak{M}$ is associated the lattice $\mathcal{L}(\mathfrak{M})$ of recursively enumerable algebraically closed substructures of $\mathfrak{M}$. (We omit the technical definitions for "recursively presented" and "algebraically closed", and proceed by example.)

EXAMPLE 1

Let $\mathfrak{M}$ be $(\omega,=)$, the set of nonnegative integers with equality. Then $\mathcal{L}(\mathfrak{M})$ is the usual Post lattice of recursively enumerable subsets of ω. We write $\mathcal{L}(\omega)$ instead of $\mathcal{L}(\mathfrak{M})$.

EXAMPLE 2

Let $\mathfrak{M}$ be the vector space V of section 2. Then $\mathcal{L}(V)$ is the lattice of recursively enumerable subspaces of V.

Decidable subspaces of V are to $\mathcal{L}(V)$ as recursive subsets of ω are to $\mathcal{L}(\omega)$. Call a C in $\mathcal{L}(V)$ creative if whenever n is an index of $W \in \mathcal{L}(V)$ and $W \cap C = \{\underline{0}\}$, we can compute an $x \in V-(W+C)$. Any two subspaces G, each generated by a creative subset of a recursive basis, differ by a recursive automorphism of V and are creative. But there are other creative subspaces. We construct $S \in \mathcal{L}(V)$ so that $V \bmod S$ is effectively homogeneous universal over all $V \bmod W$ with $W \in \mathcal{L}(V)$. Any two such S differ by a recursive automorphism of V and are creative subspaces. But there is no recursive automorphism of V mapping such a G to such an S.

Finally, we construct maximal subspaces of V. These are to $\mathcal{L}(V)$ as Friedberg's maximal sets are to $\mathcal{L}(\omega)$.

EXAMPLE 3

Let be $(Q,\leq)$, the set of rational numbers Q under order $\leq$. Then $\mathcal{L}(Q)$ is simply the lattice of all recursively enumerable subsets of Q.

Since there is a recursive 1-1 onto map on ω to Q, $\mathcal{L}(\omega)$ and $\mathcal{L}(Q)$ are effectively lattice-isomorphic. So at first it may appear no new behaviour turns up in $\mathcal{L}(Q)$. But this is misleading. We give two examples.

First, we have developed the notion of *creative* dense simple orders without endpoints – call one C. Roughly, C is a recursively enumerable subset of Q such that

i) C is dense in Q,

ii) $Q-C$ is dense in Q, and

iii) if I is any interval of Q, $C \cap I$ is a creative subset of I (uniformly). Such sets exist, and we prove any two are isomorphic by a recursive automorphism of the ordered structure $(Q,\leq)$. We produce maximal $M \in \mathcal{L}(Q)$ – i.e., M, $Q-M$ are dense but for no $X \in \mathcal{L}(Q)$ with $X \supseteq M$ are both $X-M$, $Q-X$ dense.

EXAMPLE 4

Let B be a recursively presented atomless boolean algebra. There are (for example) recursively enumerable ideals I with B mod I 'effectively universal homogeneous' over the B mod S with S recursively enumerable; and B mod I is recursively unique.

The subject goes especially smoothly in homogeneous models particularly in $\aleph_0$-categorical models like those above[1].

4. *The invariant point of view*

Klein's Erlanger program approached each geometry as the study of invariants under an appropriate transformation group. It is often profitable to view other branches of mathematics not ordinarily regarded as geometric from this point of view. If we decide which structures are to be regarded as indistinguishable in a given branch of mathematics, the important problems more readily meet the eye. Recursion theory is no exception.

A substantial part of recursion theory on ω deals with properties, relations, and operations which are invariant under the group $G_R(\omega)$ of recursive permutations of ω. This includes most of the theories of recursively enumerable sets and degrees of unsolvability. For instance the properties of simplicity, creativity, and maximality have this kind of invariance; and likewise relations such as Turing reducibility, and operations such as union and intersection.

Another part of recursion theory on ω deals with properties invariant under $S_R(\omega)$, the semigroup of 1-1 partial functions which are extendible to 1-1 partial recursive functions. This part of recursion theory includes the theory of recursive equivalence types. The properties of being effectively Dedekind (i.e., isolated) or of being indecomposable have this kind of invariance; the recursive combinatorial operators possess it too.

We may regard Post's paper [44] as the first systematic study of invariants under $G_R(\omega)$. Similarly Dekker-Myhill's monograph [60] was the first systematic study of invariants under $S_R(\omega)$.

Analogues of the classification of sets of nonnegative integers by $G_R(\omega)$ have been developed by Crossley [69], Hassett [64], Applebaum [69], Dekker [69], [71] respectively for linear orderings, groups, and vector spaces. Crossley and Dekker classified algebraic structures by isomorphisms which are recursively enumerable. Hassett and Applebaum also used isomorphisms which have 1-1 partial recursive extensions[3].

In Crossley-Nerode [74] the methods of Nerode [61] are blended with model theory to formulate this subject as the study of algebraically closed subsets of a recursively presented model $\mathfrak{M}$. These structures are classified by the category $S_R(\mathfrak{M})$ of elementary isomorphisms extendible to 1-1 partial recursive functions.

We have said that the ordinary theory of recursively enumerable sets is related to the theory of recursive equivalence types in the same way that the group $G_R(\omega)$ is related to the semigroup $S_R(\omega)$. If $G_R(\mathfrak{M})$ is the group of recursive automorphisms of $\mathfrak{M}$, it is natural to expect that there is a subject of recursively enumerable algebraically closed subsets of $\mathfrak{M}$ related to the Crossley-Nerode theory of effective isomorphism types in the same way that $G_R(\mathfrak{M})$

is related to $S_R(\mathfrak{M})$. At least this suggests that the present study of invariants under $G_R(\mathfrak{M})$ will be fruitful.

We close with a note of warning. The recursive features of algebraic constructions are particular, and differ very much from case to case. The "universal" part of this kind of investigation, which deals with general model-theoretic common features, is therefore bound to be somewhat pedestrian. But the specifics are interesting and present many challenges. This is characteristic of recursion theory.

5. *Footnotes*

[1]J. Remmel (1974 Cornell dissertation under Nerode) has developed the theory of co-recursively enumerable structures. These are algebraically closed subsets of $\mathfrak{M}$ whose complement in $\mathfrak{M}$ is recursively enumerable. This turns out to be a surprisingly rich dual to the subject described above. In a quite general context he has produced algebraically closed $\mathfrak{A} \subseteq \mathfrak{M}$ with either

i) $\mathfrak{M} - \mathfrak{A}$ hypersimple and low, or

ii) $\mathfrak{M} - \mathfrak{A}$ hypersimple and complete.

Moreover, $\mathfrak{A}$ can be chosen from any Σ_2^0 classical isomorphism type. In the vector space V over a finite field he has produced co-recursively enumerable "cohesive" spaces, both complete and incomplete.

We also remark that Remmel has lifted Hay's theory of co-recursively enumerable sets and recursive equivalence types to a theory of co-recursively enumerable algebraically closed sets and Crossley-Nerode effective isomorphism types.

[2]In 1968 Nerode raised as a test question whether all the

countable models of a complete decidable theory with only finitely many countable models could be recursively presented. In 1972 Morley gave a six model counterexample, in 1973 Peretiatkin [73] gave a three model counterexample in which exactly one of the three models has a recursive presentation. In 1974 Morley showed that if there are exactly three countable models and two have recursive presentations, then so does the third. This used similar ideas to those used by Harrington [72] to see that the prime model of differentially closed fields has a recursive presentation. T. Millar [74] (Cornell dissertation under Nerode, in preparation) has further results. The simplest are:

i) if a complete decidable theory has a recursively enumerable list of types, then the saturated model is recursively presentable;

ii) if a complete decidable theory has a Σ_2^0 list of non-principal types, these can all be omitted from some recursively presented model.

We mention another result in this type of recursive model theory, due to M. Venning [73] (unfinished Cornell dissertation under Nerode). There is a recursively presented $\aleph_0$-categorical model (in a language with finitely many relations) in which algebraic closure of finite sets is not recursive on canonical indices of finite sets.

[3]E. Eisenberg (1974 dissertation under Nerode) has shown that classification by recursively enumerable isomorphism is quite different from classification by isomorphisms extendible to 1-1 partial recursive functions. For instance, in the vector space V (over a finite field) there are Dedekind subspaces W_1, W_2 such that

i) there is an isomorphism from W_1 onto W_2 extendible to a 1-1 partial recursive function;

ii) there is no recursively enumerable isomorphism, defined on W_1 mapping W_1 onto W_2. A similar result holds for cosimple orderings. Both E. Eisenberg and C. Chell have used this method to solve a corresponding problem of Applebaum for groups, even in the cosimple case.

BIBLIOGRAPHY

APPLEBAUM, C.H.

[69] Decomposition and homomorphisms of ω-groups, Ph.D. dissertation, Rutgers University, 1969.

CROSSLEY, J.N.

[69] Constructive order types, (monograph), North-Holland, Amsterdam, 1969, 225 pp.

CROSSLEY, J.N. and A. NERODE

[74] Combinatorial Functors, (monograph), Springer-Verlag, Berlin, 1974, 144 pp. (vol. 81 of Ergebnisse series).

DEKKER, J.C.E. and J. MYHILL

[60] Recursive equivalence types, (monograph), University of California publications in mathematics, n.s. vol. 3, 1960, 67-214.

DEKKER, J.C.E.

[69] Countable vector spaces with recursive operations, Part I, J. Symbolic Logic 34, 1969, 363-387.

[71] Countable vector spaces with recursive operations, Part II, J. Symbolic Logic 36, 1971, 477-495.

EISENBERG, E.

[74] Effective isomorphisms of groups and other structures, Ph.D. dissertation, Cornell University, 1974.

HARRINGTON, L.

[72] Recursively presentable prime models, 10 pp. (mimeo, 1972).

HASSETT, M.J.

[64] Some theorems on regressive isols and isolic groups, Ph.D. dissertation, Rutgers University, 1964.

LACHLAN, A.H.

[70] On some games which are relevant to the theory of recursively enumerable sets, Ann. Math. 91, 1970, 291-310.

NERODE, A.

[61] Extensions to isols, Ann. Math. 73, 1961 , 362-403.

PERETIATKIN, M.G.

[73] Example of a complete decidable theory which has exactly three denumerable models, but only one of them is recursive, (mimeo, 1973).

POST, E.

[44] Recursively enumerable sets of positive integers and their decision problems, Bull. Amer. Math. Soc. 50, 1944, 284-316.

REMMEL, J.

[74] Co-re structures, Ph.D. dissertation, Cornell University, 1974.

YATES, M.

[74] A new approach to degree theory, (monograph), (mimeo, 1974).

* * *

Department of Mathematics, University of Rochester, Rochester, New York

Department of Mathematics, Cornell University, Ithaca, New York.

AN EXPOSITION OF FORCING

A. Mostowski

P.J. Cohen invented, in 1963, a method of constructing models for Zermelo-Fraenkel set theory, hereafter abbreviated ZF. In the present lectures we shall describe this method with some modifications (due mainly to Solovay) and apply it to a proof that the continuum hypothesis is independent of ZF. The independence proof is taken over from Cohen [66] without change.

I wish to thank Dr. W. Guzicki for his helpful discussions on the topic of these lectures and for having shown me his notes of similar lectures delivered in the University of Nijmegen.

1. *Logical preliminaries*

The language of ZF is the first order language with identity and with one binary predicate ε. We shall denote this language by L. We write $x \varepsilon y$ instead of $\varepsilon(x,y)$. The logical symbols we use are: $\neg$, $\rightarrow$, $\equiv$, &, $\vee$, $\forall x$, $\exists x$. We abbreviate $\forall x[x \varepsilon a \rightarrow F]$ by $\forall x \varepsilon a\, F$ and $\exists x\, [x \varepsilon a \,\&\, F]$ by $\exists x \varepsilon a\, F$. Moreover we write $x \subseteq y$ instead of $\forall x \varepsilon z (z \varepsilon y)$ and $x \subset y$ instead of $x \subseteq y \,\&\, x \neq y$. The same abbreviations are used for other variables also.

If ϕ is a formula of L then $Fr(\phi)$ denotes the set of the free variables of ϕ.

The axioms of ZF are as follows:

1. Axiom of extensionality: $\forall x \forall y [x = y \equiv \forall z (z \varepsilon x \equiv z \varepsilon y)]$,
2. Existence of pairs: $\forall x \forall y \exists z \forall t [t \varepsilon z \equiv (t = x \vee t = y)]$,
3. Existence of unions: $\forall x \exists y \forall z [z \varepsilon y \equiv \exists t \varepsilon x (z \varepsilon t)]$,
4. Existence of power sets: $\forall x \exists y \forall z [z \varepsilon y \equiv z \subseteq x]$,
5. Existence of infinite sets: $\exists x \exists y \varepsilon x \forall z \varepsilon x \exists t \varepsilon x [z \subset t]$,
6. Axiom of foundation: $\forall x \forall y \varepsilon x \exists z \varepsilon x \ \forall t \varepsilon x [\neg(t \varepsilon z)]$,

7_ϕ. Axiom scheme of comprehension: $\forall x \exists y \forall z [z \varepsilon y \equiv (\phi \,\&\, z \varepsilon x)]$,

8_ψ. Axiom scheme of replacement: $\forall x \exists y \forall z \varepsilon x \ \forall t [\psi \rightarrow \exists t \varepsilon y\, \psi]$.

In 7_ϕ, ϕ can be any formula of L in which the variable z but not y is free; in 8_ψ, ψ can be any formula of L in which the variables z, t are free but the variable y is not free.

The axiom of choice is the following sentence:

$$\text{AC.} \quad \forall x \forall y \varepsilon x \forall z \varepsilon x \exists s \varepsilon y \forall t [\neg(t \varepsilon y \,\&\, t \varepsilon z) \vee z = y] \rightarrow \exists w \forall y \varepsilon x \exists v \varepsilon y \forall t \varepsilon y (t \varepsilon w \equiv t = v).$$

We shall denote by ZFC the system obtained by adjoining AC to ZF.

Our meta-theory in which we shall study models of ZF will be the set theory ZFC enriched by one additional axiom SM due to Cohen. We shall formulate this axiom below after introducing some definitions. We shall freely use the current set-theoretical notation and shall write set-theoretical formulae using the same logical symbols as in the language L. The membership relation however will be denoted by $\in$.

A family F of sets is called *transitive* if $x \in y \in F$ implies $x \in F$.

We shall assume as known the concept of a relational system and the notion of satisfaction of a formula in such systems. We use the customary notation $M \models \phi[a]$ for: ϕ is satisfied in M by the assignment a of elements of M to the free variables of ϕ. We shall use the same notation also for languages arising from L by adjunction of constants and additional predicates.

Relational systems in which the universe is a transitive family of sets and ε is interpreted as the membership relation will be called *models*.

We shall denote by the same letter a relational system M and its universe. The set of all assignments of elements of M to the free variables of a formula ϕ can then be denoted by $M^{Fr(\phi)}$.

LEMMA 1.1

If M *is a transitive family of sets such that the set* $\omega = \{\emptyset,\{\emptyset\},\{\emptyset,\{\emptyset\}\},\dots\}$ *belongs to* M *and* M *is closed with respect to the formation of pairs and unions then axioms* 1,2,3,5 *and* 6 *are valid in* M.

Proof: routine.

The existence of a model M in which the remaining axioms of ZF are valid cannot be proved on the basis of the usual axioms of set theory even if we assume an additional axiom stating the consistency of ZF.

The additional axiom SM which we mentioned above states:

(SM) There is a denumerable model of ZF.

REMARKS

(i) In view of the Skolem-Löwenheim theorem (SM) follows from a weaker axiom: there is a model of ZF.

(ii) From Gödel [40] it follows that each model M of ZF contains a submodel M' such that all the axioms of ZFC are valid in M'. Thus (SM) can also be replaced by the axiom: there is a model of ZFC.

2. *Algebraic preliminaries*

Let P be a partially ordered set. We denote by $\leq$ the ordering relation. The elements of P are called "conditions" and are denoted by letters $p, q, r, \ldots$. If $p \leq q$ then p is called an *extension* of q; if moreover $p \neq q$ then we write $p < q$ and call p a *proper* extension of q. If $\exists r\, [r \leq p \;\&\; r \leq q]$ then p and q are called *compatible*, otherwise incompatible.

Assumptions concerning P.

1. $\leq$ is a partial ordering.
2. Each condition has a proper extension.
3. If $p \nleq q$ then there is an r such that $r \leq p$ and r is incompatible with q.

Partial orderings satisfying 3 are called *separable*.

Filters.

A set $F \subseteq P$ is called a filter if

(i) $F \neq \emptyset$,

(ii) $p \leq q$ and $p \in F$ imply $q \in F$,

(iii) $p \in F$ and $q \in F$ imply $\exists r \varepsilon F\ [r \leq p\ \&\ r \leq q]$.

LEMMA 2.1

Each filter can be extended to a maximal filter.

Proof. Use Zorn's lemma and the remark that the union of a chain of filters is a filter.

LEMMA 2.2

If F *is a filter and* p *is incompatible with an element of* F *then there is a maximal filter* F' *such that* $p \notin F'$ *and* $F \subseteq F'$.

Proof. Use Zorn's lemma to obtain a maximal element F' of the family of filters containing F as a subset and not containing p as an element.

LEMMA 2.3

If $q \nleq p$ *then there is a maximal filter containing* q *but not* p.

Proof. By separability there is an r in P such that $r \leq q$ and r is incompatible with p. By 2.2 the filter $\{x \in P : x \geq r\}$ can be extended to a maximal filter and this filter satisfies 2.3.

LEMMA 2.4

If $p_0 \geq p_1 \geq \ldots$ *then there is a maximal filter containing all the* p_n.

Proof. Extend the filter $\{x \in P : \exists n\ (x \geq p_n)\}$ to a maximal one.

We shall now define a topological space. Let $\mathcal{X}$, or more precisely $\mathcal{X}(P,\leq)$, be the set of all maximal filters of P and take as an open sub-basis of $\mathcal{X}$ the family of all sets

$$[p] = \{F \in \mathcal{X} : p \in F\}.$$

Sets of the form $[p]$ will be called *neighbourhoods*.

THEOREM 2.5

$\mathcal{X}$ is a topological space which satisfies the Baire category theorem.

Proof. Open sets are defined as arbitrary unions of neighbourhoods $[p]$. Hence arbitrary unions of open sets are open.

The intersection of two open sets $\mathcal{G}_1, \mathcal{G}_2$ is open. For if $F \in \mathcal{G}_1 \cap \mathcal{G}_2$ then there are p_1, p_2 such that $F \in [p_1] \subseteq \mathcal{G}_1$ and $F \in [p_2] \subseteq \mathcal{G}_2$. Hence $p_1 \in F$ and $p_2 \in F$ and therefore there is an r in F such that $r \leq p_1$ and $r \leq p_2$. It follows that $F \in [r]$ and $[r] \subseteq [p_1] \cap [p_2] \subseteq \mathcal{G}_1 \cap \mathcal{G}_2$. Thus $\mathcal{G}_1 \cap \mathcal{G}_2$ is either void or is a union of neighbourhoods. Hence $\mathcal{X}$ is a topological space.

We now prove that if $\mathcal{G}_n \subseteq \mathcal{X}$ and $\mathcal{G}_n$ is open and dense in $\mathcal{X}$ for $n = 1,2,3,\ldots$ then $\bigcap_n \mathcal{G}_n \neq \emptyset$ (Baire theorem).

Put $G_n = \{p \in P : [p] \subseteq \mathcal{G}_n\}$. From the density of $\mathcal{G}_n$ it follows that for each q in P there is a p in G_n such that $p \leq q$, i.e. that G_n is dense in P. Let p_0 be arbitrary. From the density of G_1 we infer that $p_1 \leq p_0$ for some p_1 in G_1. From the density of G_2 we infer that $p_2 \leq p_1$ for some p_2 in G_2.

Continuing in this way we obtain a decreasing sequence $p_0 \geq p_1 \geq \dots$. By 2.4 there is a maximal filter F such that $p_n \in F$ for each n and it follows that $F \in [p_n] \subseteq \mathcal{G}_n$ for each n, i.e. $F \in \bigcap_n \mathcal{G}_n$.

3. *Heuristic remarks about models in which the continuum hypothesis fails*

We assume that we live in a world in which the continuum hypothesis (abbreviated CH) is true and want to construct another world in which CH is false, i.e. in which there is an injection f of ω_α into 2^ω for some $\alpha > 1$. No such injection exists in our world but we can say that if it exists anywhere, we have in our world its finite approximations. These approximations are finite functions p with domains $\subseteq \omega_\alpha \times \omega$ and range $\subseteq \{0,1\}$. Indeed, if f is our hypothetical injection and D is a finite subset of $\omega_\alpha \times \omega$ then putting on D $p(\xi,n) = f(\xi)(n)$ (the value of $f(\xi)$ for the argument n) we obtain a finite approximation of f which, because of the finiteness of D, belongs to our world. Mappings of finite subsets of $\omega_\alpha \times \omega$ into $\{0,1\}$ form a set P which is partially ordered by the relation of inverse inclusion : $p \leq q$ means that $p \supseteq q$. Each maximal filter of P determines a function defined on the whole set $\omega_\alpha \times \omega$; conversely each mapping f of $\omega_\alpha \times \omega$ into $\{0,1\}$ determines a maximal filter of P consisting of all the finite approximations of f. Maximal filters which belong to our world determine functions which are not one-to-one. It will be our task to find an extension of our world with new maximal filters of P.

Two methods are known at present to construct such extensions. One of them identifies sets with two-valued functions whose values are arbitrary elements of suitable Boolean algebras. Another method identifies the "world" with a denumerable model M of ZF and extends M by adding to it certain subsets of P not previously contained in M.

Below we shall sketch this second method.

4. *Constructible sets*

Let M be a denumerable model of ZF and let $On_M = On \cap M$ be the set of the ordinals of M. We denote by $rk(x)$ the rank of x defined by induction as follows

$$rk(x) = \sup\{rk(y) : y \in x\}.$$

The set $\{x \in M : rk(x) \leqslant \alpha\}$ will be denoted by M_α.

Let X be a subset of M whose rank α_0 belongs to On_M and put $\Lambda_\alpha = \emptyset$ or $\Lambda_\alpha = \{X\}$ according as $\alpha < \alpha_0$ or $\alpha \geqslant \alpha_0$. We are going to define a family $M[X]$ such that $M \subseteq M[X]$ and $X \in M[X]$. For suitable X this family will be the required model.

For each family B of sets, each formula ϕ of L such that $v \in Fr(\phi)$ and each assignment $\gamma \in B^{Fr(\phi)-\{v\}}$ we define the *extension* of ϕ in B with respect to γ as

$$E_{\phi,\gamma,B} = \{x \in B : B \models \phi[x,\gamma]\}$$

where x,γ is an assignment which correlates x to v and is otherwise identical with γ.

The family of all sets $E_{\phi,\gamma,B}$ where ϕ ranges over formulae of L such that $v \in Fr(\phi)$ and γ ranges over $B^{Fr(\phi)-\{v\}}$ is called the *derived family* and will be denoted by B'.

We define now $M[X]$ as the union $\cup\{B_\alpha[X] : \alpha \in On_M\}$ where $B_0[X] = \emptyset$, $B_\lambda[X] = \cup\{B_\alpha[X] : \alpha < \lambda\}$ if λ is a limit number and $B_{\alpha+1}[X] = B'_\alpha[X] \cup M_\alpha \cup \Lambda_\alpha$

Elements of $M[X]$ will be called *constructible in* X. The following properties of the families $B_\alpha[X]$ and $M[X]$ are easy to prove.

LEMMA 4.1

$M \subseteq M[X]$.

Proof. If $m \in M$, then $m \in M_\alpha$ where $\alpha = rk(m)$ and hence $m \in B_{\alpha+1}[X]$.

LEMMA 4.2

$X \in M[X]$.

Proof. $X \in \Lambda_{\alpha_0}$ and hence $X \in B_{\alpha_0+1}[X]$.

LEMMA 4.3

$B_\alpha[X]$ *is transitive for each* $\alpha \in On_M$.

Proof. We use induction. The lemma is true for $\alpha = 0$. If it is true for $\alpha < \lambda$ where λ is a limit number then it is true for λ. If $x \in y \in B_{\alpha+1}[X]$ then either $y \in B'_\alpha[X]$ or $y \in M_\alpha$ or $y = X$. In the first case y is the extension of a formula in $B_\alpha[X]$ and hence $x \in B_\alpha[X]$. In the second case $rk(y) \leq \alpha$ and hence $rk(x) < \alpha$ and thus $x \in M_{rk(x)} \subseteq B_{\alpha+1}[X]$. The last case can occur only if $\alpha \geq \alpha_0$. We have then $x \in X$, $rk(x) < \alpha$ and $x \in M$ because $X \subseteq M$. Hence $x \in M_\alpha$ and $x \in B_{\alpha+1}[X]$.

LEMMA 4.4

$B_\alpha[X] \subseteq B_\beta[X]$ *whenever* $\alpha \leq \beta$.

Proof. It is sufficient to prove the lemma for the case $\beta = \alpha + 1$. Let $a \in B_\alpha[X]$ and consider the formula $v \in w$ and the assignment γ such that $\gamma(w) = a$. The extension of $v \in w$ in $B_\alpha[X]$ with respect to γ is $\{x \in B_\alpha[X] : x \in a\} = a \cap B_\alpha[X] = a$ because all elements of a belong to $B_\alpha[X]$ in view of the transitivity of $B_\alpha[X]$. Hence $a \in B_{\alpha+1}[X]$.

LEMMA 4.5

$M[X]$ *is transitive.*

Proof. The union of a chain of transitive sets is transitive.

LEMMA 4.6

$B_\alpha[X] \in B_{\alpha+1}[X]$.

Proof. $B_\alpha[X]$ is the extension of the formula $v = v$ in $B_\alpha[X]$ with respect to the void valuation.

LEMMA 4.7

$B_\alpha[X] \in M[X]$ *for each* $\alpha < On_M$.

Proof. By 4.6.

LEMMA 4.8

If $x \in B_\alpha[X]$ *then* $rk(x) < \alpha$.

Proof. For $\alpha = 0$ the lemma is obvious. Assume that $\alpha > 0$ and the lemma holds for $\xi < \alpha$. If α is a limit number there is nothing to prove. If $\alpha = \xi + 1$ and $x \in B_{\xi+1}[X]$ then x is a set of elements which belong to $B_\xi[X]$, hence their ranks are $< \xi$ and the lemma follows.

LEMMA 4.9

$M[X]$ is closed under the formation of pairs and unions.

Proof. Let $a, b \in M[X]$ and let α be an ordinal such that $a, b \in B_\alpha[X]$. The extension of the formula $v = w \vee v = u$ in $B_\alpha[X]$ with respect to the assignment $\gamma(w) = a$, $\gamma(u) = b$ is $\{a,b\}$. The extension of the formula $\exists u\,(u \in w \,\&\, v \in u)$ in $B_\alpha[X]$ with respect to the assignment $\gamma(w) = a$ is $\cup a$. Thus $\{a,b\}$ and $\cup a$ are elements of $B_{\alpha+1}[X]$.

LEMMA 4.10

Axioms 1, 2, 3, 5, 6 of ZF are valid in $M[X]$.

Proof. This follows from Lemmas 1.1, 4.5, 4.9 and the remark that the set ω referred to in Lemma 1.1 is an element of M and hence of $M[X]$.

LEMMA 4.11

The ordinals of $M[X]$ are the same as the ordinals of M.

Proof. x is an ordinal of $M[X]$ if and only if it is a transitive set which belongs to $M[X]$ and whose elements are transitive sets. A transitive set all of whose elements are transitive sets is equal to its rank. By 4.8 the rank of an element of $M[X]$ is $< On_M$. It follows that each ordinal of $M[X]$ is an ordinal of M. The converse implication is obvious.

5. *The ramified language* RL

The important feature of the technique invented by Cohen is that it allows us to speak about the model $M[X]$ remaining so to speak in the model M (we speak in "our world" M about the "fictitious world" which extends M). This is due to the fact that each element of M[X] has a "name" in M. These "names" will be expressions of an auxiliary language which we shall call the ramified language or RL for short. The expressions of RL will be elements of M.

The language RL has an infinite number of variables $v_0, v_1, \ldots,$ two binary predicates ε, $\approx$ called the membership and identity predicates, propositional connectives $\neg$, & and the universal quantifier $\forall$. Besides these expressions the language RL will have infinitely many constants and infinitely many one-place predicates which will be described a little later.

The rules of formation will be the usual ones. Thus constants and variables are terms of RL. If t_1 and t_2 are terms and V is a one-place predicate then $t_1 \varepsilon t_2$, $t_1 \approx t_2$, Vt_1 are atomic formulae. If ϕ and ψ are formulae, then so are (ϕ) & (ψ), $\neg(\phi)$ and $\forall v\ (\phi)$ for each variable v. The distinction between free and bound occurrences of a variable in a formula is assumed as known; $Fr(\phi)$ denotes the set of all free variables of ϕ. If γ is a sequence of constants and $Dom(\gamma) \subseteq Fr(\phi)$ then $\phi(\gamma)$ denotes the formula resulting from ϕ by substituting $\gamma(v)$ for v throughout ϕ for each v in $Dom(\gamma)$.

Writing formulae of RL we shall often use connectives other than & and $\neg$ and also the existential quantifier. These symbols are then thought of as abbreviations. Also we shall often use letters v, w, u, etc. instead of $v_0, v_1, v_2, \ldots$.

Since we want to treat expressions of RL as elements of M we identify the primitive symbols of RL with certain elements of M and agree that if an expression is obtained by writing the symbols $A, B, C, \ldots, H$ one after another, then the whole expression is to be identified with the sequence $\langle A,B,C,\ldots,H\rangle$.

We identify v_j with $\langle 0,j\rangle$ and the symbols ε, $\approx$, $\&$, $\neg$, (,), $\forall$ with the pairs $\langle 1,j\rangle$, $j = 0,1,\ldots,5,6$. Elements of M which will serve as constants and as one-place predicates will be defined later.

We note that from now on L will be treated as a part of RL.

We describe now the additional predicates and constants of RL. For each ordinal $\alpha \in On_M$ we have in RL a one-place predicate V_α which we shall identify with the pair $\langle 2,\alpha\rangle$. Intuitively V_α denotes the set $B_\alpha[X]$.

We put $\underline{m} = \langle 3,m\rangle$ for each m in M and $\sigma = \langle 4,0\rangle$. For α in On_M, for ϕ a formula of L such that $v \in Fr(\phi)$ and each sequence γ with domain $Dom(\gamma) = Fr(\phi) - \{v\}$ we put

$$c_{\alpha,\phi,\gamma} = \langle 5,\alpha,\phi,\gamma\rangle .$$

Constants of RL can now be defined by transfinite induction:

$$C_0 = \emptyset,\ C_\lambda = \cup\{C_\alpha : \alpha < \lambda\} \text{ for limit numbers } \lambda,$$

$$C_{\alpha+1} = C_\alpha \cup \{\underline{m} : rk(m) \leqslant \alpha\} \cup \Gamma_\alpha \cup C'_\alpha$$

where $\Gamma_\alpha = \emptyset$ or $\Gamma_\alpha = \{\sigma\}$ according as $\alpha < \alpha_0$ or $\alpha \geqslant \alpha_0$ and C'_α is the set of all $c_{\alpha,\phi,\gamma}$ where ϕ is a formula of L, $v \in Fr(\phi)$ and $\gamma \in C_\alpha^{\,Fr(\phi)-\{v\}}$.

Let $C = \cup\{C_\alpha : \alpha \in On_M\}$. Elements of C are called constants of RL. For each c in C we define its order $\rho(c)$ as the least α such that $c \in C_\alpha$. Thus $\rho(c)$ is always a successor ordinal.

The intuitive meaning of the constants is as follows: $\underline{m}$ denotes m, σ denotes X and $c_{\alpha,\phi,\gamma}$ denotes the extension of ϕ in $B_\alpha[X]$ with respect to the assignment which correlates with each variable $w \in Fr(\phi)$ different from v the object denoted by the constant $\gamma(w)$.

A formula of RL in which each quantifier $\forall x$ (where x is any variable) is followed by an expression of the form $(V_\alpha x \rightarrow \psi)$ is called a limited formula. The index α may differ from quantifier to quantifier within the formula. We shall abbreviate $\forall x[V_\alpha x \rightarrow \psi]$ by $\forall_\alpha x\ \psi$ and $\exists x\ [V_\alpha x\ \&\ \psi]$ as $\exists_\alpha x\ \psi$. The order $\rho(\phi)$ of a limited formula is defined as the larger of the following two ordinals: $\max\{\rho(c) : c$ occurs in $\phi\}$, $\max\{\alpha : V_\alpha$ occurs in $\phi\}$.

Several times we shall have occasion to use an ordering $\prec$ of limited sentences defined as follows: $\phi \prec \psi$ if and only if one of the following conditions is satisfied:

1. $\rho(\phi) < \rho(\psi)$;
2. $\rho(\phi) = \rho(\psi)$ and ϕ contains fewer occurrences of logical operations (i.e. connectives and quantifiers) than ψ;
3. ϕ and ψ are atomic sentences, $\rho(\phi) = \rho(\psi)$ and ϕ has the form $c_1 \varepsilon c_2$ with $\rho(c_1) < \rho(c_2)$ whereas ψ does not have this form;
4. ϕ and ψ are atomic sentences, $\rho(\phi) = \rho(\psi)$, ϕ has the form $c_1 \approx c_2$ and ψ has the form $c_3\ \varepsilon\ c_4$ with $\rho(c_3) \geq \rho(c_4)$.
5. ϕ and ψ are atomic sentences, $\rho(\phi) = \rho(\psi)$, ϕ has the form $c_1 \approx c_2$ where $\rho(c_1) \geq \rho(c_2)$ and ψ has the form $V_\alpha c_1$.

If ϕ is a formula, $\alpha \in On_M$, x is a variable and ψ arises from ϕ by replacing an occurrence of the quantifier $\forall x$ by $\forall x\ [V_\alpha x \to V_\alpha c_1]$ then ψ is said to arise from ϕ by a relativization of the considered occurrence of $\forall x$ to V_α. We shall denote by $\phi^{(\alpha)}$ a formula which arises from ϕ by relativizing all occurrences of the quantifier to V_α.

LEMMA 5.1

$C_\alpha \in M$ for each $\alpha \in On_M$.

This follows from the remark that the operation which yields C_α from the sequence $\{C_\xi\}_{\xi<\alpha}$ is definable in M.

LEMMA 5.2

For each $\alpha \in On_M$ the set of all formulae of order α belongs to M.

This is so because such formulae are finite sequences built according to recursive rules from variables and symbols $\neg$, ε, $\approx$, V_ξ, $\forall$, ε, c where $\xi < \alpha$ and $c \in C_\alpha$. Since all these symbols form a set which belongs to M, so do all their finite sequences. In view of the recursive character of the formation rules, the formulae themselves also form a set in M.

In connection with these lemmas it is worth noting that neither C nor the set of all formulae of RL belong to M, although of course they are definable subsets of M.

LEMMA 5.3

The relation $\prec$ is a definable partial well ordering of the limited sentences. Its restriction to the set of sentences of order $< \alpha$ is a set in M for each $\alpha \in On_M$.

Proof. The second part follows from the first and Lemma 5.2. The definability of $\prec$ follows from the remark that the function ρ and clauses 1-4 of the definition of $\prec$ are definable in M. The well-foundedness of $\prec$ is proved as follows: Let A be a non void set of limited sentences and A_0 its subset consisting of sentences of a possibly small order. If A_0 does not contain atomic sentences of the form $c_1 \varepsilon c_2$ or $c_1' \approx c_2'$ then no element of A_0 has a predecessor in A. If A_0 contains at least one sentence of the form $c_1' \approx c_2'$ but no sentence of the form $c_1 \varepsilon c_2$ with $\rho(c_1) < \rho(c_2)$ then each sentence $c_1' \approx c_2'$ which belongs to A_0 is a minimal element of A. Finally if A_0 contains at least one sentence $c_1 \varepsilon c_2$ with $\rho(c_1) < \rho(c_2)$ then each such sentence is a minimal element of A_0.

Limited sentences ϕ which have no predecessors with respect to $\prec$ are $\underline{0} \varepsilon \underline{0}$, $\underline{0} \approx \underline{0}$ and $V_0\underline{0}$.

Values of the constants. Let X be a subset of a set which belongs to M and has rank α_0. For each constant c we define by induction the value of c for the argument X; we denote this value by $c^*[X]$.

Let us assume that $\alpha \in On_M$ and that $c^*[X]$ is already defined for c in C_β with $\beta < \alpha$. Let $c \in C_\alpha$. If α is a limit number there is nothing to define. If $\alpha = \beta + 1$ there are three possibilities:

1) $c \in C'_\beta$, 2) $c = \underline{m}$ where $rk(m) \leqslant \beta$, 3) $c = \sigma$.

In cases 2) and 3) we put $c^*[X] = m$ and $c^*[X] = X$ respectively. In case 1) $c = c_{\beta,\phi,\gamma}$ where ϕ is a formula of L, $v \in Fr(\phi)$ and

$\gamma \in C_\beta^{Fr(\phi)-\{v\}}$. In this case we put $c^*[X] = E_{\phi,\gamma^*[X],B_\beta[X]}$ where $\gamma^*[X]$ is a sequence with domain $Fr(\phi) - \{v\}$ whose terms are values of the terms of γ.

LEMMA 5.4

For each $\alpha \in On_M$ *the set* $B_\alpha[X]$ *is identical with the set of all* $c^*[X]$ *where* c *ranges over* C_α.

Proof by an obvious induction.

Lemma 5.4 says that each element of $M[X]$ has a name in C. As a matter of fact each element of $M[X]$ has many names; given an element x in $M[X]$ the set of all its names is a subset of M but not an element of M.

It can be shown that the function which correlates $c^*[X]$ with c is definable in $M[X]$. This will follow from theorems which will be established in the next section and from the following weaker result:

LEMMA 5.5

Let N *be a model of* ZF *such that* $N \supseteq M$ *and let* f *be a function of two arguments the first of which ranges over* C *and the second over subsets belonging to* N *of a fixed set* $P \in M$ *of rank* α_0 *and which is defined by the equation* $f(c,X) = c^*[X]$. *Then* f *is definable in* N.

Proof. f is defined by transfinite induction on the order of c. Since the theorem on definitions by transfinite induction is valid in N, we obtain the desired result.

We close our discussion with a remark on semantics of the language RL. The relational structures which can serve as models for formulae of RL have an infinite type because there are infinitely many constants in RL and also infinitely many one-place predicates in addition to the two binary ones ε and $\approx$. Apart from that the model theory of the language RL does not differ from the model theory of any first order language.

In most cases we shall deal with models whose universe is $M[X]$ and where the predicates $\varepsilon, \approx, V_\alpha$ are interpreted as $\in, =, B_\alpha[X]$ and the constants c as $c^*[X]$. Such a model will be denoted briefly by $M[X]$.

6. *Reduction of properties of* M[X] *to* M

The following example shows that, in general, $M[X]$ is not a model of ZF : Let X be a subset of $\omega \times \omega$ such that the relation mXn orders ω similarly to On_M. The existence of X follows from our assumption that M and hence also On_M are denumerable. The family $On \cap M[X]$ is equal to On_M because each element of $M[X]$ has a rank $< On_M$ (see Lemma 4.11). On the other hand $X \in M[X]$ and X has the order type On_M and so the theorem: "for each well ordering there is a similar ordinal" is not valid in $M[X]$. Yet this theorem is provable in ZF. Hence not all axioms of ZF are valid in $M[X]$.

There are no general criteria which would allow us to decide for which sets X the family $M[X]$ is a model of ZF. A sufficient condition which we shall discuss below says that the satisfaction relation

$$M[X] \models \phi[\gamma^*[X]] \tag{1}$$

be expressible in M. More exactly we require that (1) be equivalent to the fact that a formula depending on ϕ be satisfied in M by γ (i.e. the sequence of names of the terms of $\gamma^*[X]$) and by an element of X.

To return to the intuitive picture given in Section 3 where X was a maximal filter in P we may say that we require (1) to be equivalent to a relation definable in M holding between names of terms of $\gamma^*[X]$ and an approximation of an object determined by X.

The exact definition is as follows:

DEFINITION

We say that X is reducible to M if for each formula ϕ of L there is a formula Φ_ϕ with one more free variable such that for each $\gamma \in C^{Fr(\phi)}$

(2) $$M[X] \models \phi[\gamma^*[X]] \equiv \exists x \varepsilon X\ [M \models \Phi_\phi[x,\gamma]].$$

Instead of $M \models \Phi_\phi[x,\gamma]$ we shall say that the condition x establishes ϕ at γ in M[X]. (The term in general use is: x forces $\phi(\gamma)$. We selected another term in order to reserve the word "forcing" for the case of a particular formula Φ_ϕ.)

To derive properties of M[X] where X is reducible to M we need the notion of a normal function. Such a function is a strictly increasing and continuous mapping $f : On_M \to On_M$. We call α a critical number for f if $f(\alpha) = \alpha$.

LEMMA 6.1

If f *is a normal function which is definable in* M *then for each* α_0 *in* On_M *there is a critical number* α *of* f *such that* $\alpha_0 < \alpha \in On_M$.

Proof. Define a sequence $\{\alpha_n\}$ by induction on $n : \alpha_{n+1} = f(\alpha_n)$. This sequence belongs to M since the theorem on inductive definitions is valid in M. It follows that $\sup\{\alpha_n\} \in M$ and we easily prove that this supremum is the required critical number.

Another important auxiliary result is the following

THEOREM 6.2 (the reflection theorem)

Let X *be reducible to* M *and let* ϕ *be a formula of* L. *There exists a normal function* f_ϕ *definable in* M *such that whenever* α *is a critical number of* f_ϕ *and* $\gamma \in C_\alpha^{Fr(\phi)}$ *then*

$$M[X] \models \phi[\gamma^*[X]] \equiv B_\alpha[X] \models \phi[\gamma^*[X]].$$

REMARK

The reflection theorem is really a theorem scheme: for each formula ϕ we construct separately a definition of a normal function f_ϕ.

Proof. If ϕ has no quantifiers then we take for f_ϕ the identity function $f_\phi(x) = x$. If f_ϕ, f_ψ are already defined and $\theta = \neg\phi$, $\zeta = \phi \,\&\, \psi$ then we take $f_\theta = f_\phi$, $f_\zeta = f_\phi \circ f_\psi$. Using the fact that a superposition $f \circ g$ of two normal definable functions is again such a function and each of its critical numbers is a critical number of the functions f and g we convince ourselves easily that if the reflection theorem is valid for the formulae ϕ, ψ then it is also valid for the formulae $\neg\phi$ and $\phi \,\&\, \psi$. The theorem is also trivially valid for the formula $\exists v\, \phi$ if $v \notin Fr(\phi)$.

We shall now prove the theorem for the formula $\exists v\ \phi$ where $v \in Fr(\phi)$.

For each $\gamma \in C^{Fr(\phi)-\{v\}}$ we put

$$g(\gamma) = \sup_{p\in P}\{\min_{\eta\in On_M} \{\exists c\varepsilon C_\eta\ M \models \Phi_\phi[p,(c,\gamma)]\}\}.$$

Thus $g(\gamma)$ is the least ordinal λ with the property that for each condition p if there is a constant c such that p establishes ϕ at (c,γ) in $M[X]$ then there is a constant c already in C_λ such that the same p establishes ϕ at (c,γ) in M.

Using g we define a function h by transfinite induction:

$h(0) = 0$, $h(\lambda) = \sup\{h(\alpha) : \alpha < \lambda\}$ if λ is a limit number,

$h(\alpha+1) = \max(h(\alpha) + 1, \sup\{g(\gamma) : \gamma \in C_\alpha^{Fr(\phi) - \{v\}}\})$.

h is of course definable in M since the theorem on inductive definitions is valid in M. Moreover h is strictly increasing and continuous i.e. h is normal.

It follows from the last clause of the definition that $h(\alpha+1)$ is an upper bound of the values of $g(\gamma)$ for γ ranging over $C_\alpha^{Fr(\phi)-\{v\}}$. Hence if λ is a limit number and $\gamma \in C_\lambda^{Fr(\phi)-\{v\}}$ then $g(\gamma) < h(\lambda)$ because there are only finitely many terms of γ and each of them belongs to a C_α, with $\alpha < \lambda$; hence there is an $\alpha < \lambda$ such that all terms of γ belong to C_α and so $g(\gamma) < h(\alpha+1) < h(\lambda)$ because h is strictly increasing.

Now let f be the normal function $f_\phi \circ h$ and let α be one of its critical numbers. We shall show that if $\gamma \in C_\alpha^{Fr(\phi)-\{v\}}$ then

(3) $\qquad M[X] \models \exists v\ \phi[\gamma^*[X]] \equiv B_\alpha[X] \models \exists v\ \phi[\gamma^*[X]]$.

First assume the right-hand side of this equivalence. Hence there is an element x of $B_\alpha[X]$ which together with $\gamma^*[X]$ satisfies ϕ in $B_\alpha[X]$. Let $c \in C_\alpha$ be a name of x. Using the inductive assumption and the remark that α is a critical number of f_ϕ we obtain the left-hand side of (3).

Now we assume the left-hand side of (3). There is thus an element of $M[X]$ which, together with $\gamma^*[X]$ satisfies ϕ in $M[X]$. This element can be represented as $c^*[X]$ where c is a constant. Since we assume that X is reducible we obtain a condition p in X which establishes ϕ at (c,γ) in $M[X]$. Using the definition of g we infer that there is a constant c_1 in $C_{g(\gamma)}$ such that p establishes ϕ at (c_1,γ) in $M[X]$. Since α is a critical number of f_ϕ we obtain $M[X] \models \phi[c_1^*[X], \gamma^*[X]]$ and since $g(\gamma) < h(\alpha) = \alpha$ we obtain $(c,\gamma) \in C_\alpha^{Fr(\phi)}$. In view of the inductive assumption we obtain therefore the right-hand side of (3).

We can now verify the validity of Axioms 4, 7 and 8 in $M[X]$.

THEOREM 6.3

If X *is reducible to* M *then the axiom of comprehension is valid in* $M[X]$.

Proof. Let ϕ be a formula of L such that $v \in Fr(\phi)$. Let $a \in M[X]$ and $g \in M[X]^{Fr(\phi)-\{v\}}$. We have to show that there is a b in $M[X]$ such that for each x in $M[X]$

(4) $$x \in b \equiv x \in a \;\&\; M[X] \models \phi[(x,g)].$$

We can represent a as $c^*[X]$ and g as $\gamma^*[X]$ where $c \in C$ and $\gamma \in C^{Fr(\phi)-\{v\}}$. Let α be an ordinal such that c and all the terms of γ be elements of C_α. Using the reflection theorem we find an ordinal $\beta \geq \alpha$ such that for each sequence $(c_1, \gamma_1) \in C_\beta^{Fr(\phi)}$

$$M[X] \models \phi[c_1^*[X], \gamma_1^*[X]] \equiv B_\beta[X] \models \phi[c_1^*[X], \gamma_1^*[X]].$$

Let us consider the formula $\bar{\phi} = \phi \,\&\, (v \in w)$ where w is a new variable and let $\bar{\gamma}$ be an assignment which coincides with γ on the free variables of $\bar{\phi}$ the variable w excepted and correlates c with w. We claim that the set $b = c^*_{\beta,\bar{\phi},\bar{\gamma}}[X]$ satisfies (4). By definition b is the extension of $\bar{\phi}$ in $B_\beta[X]$ with respect to $\bar{\gamma}^*[X]$ i.e.

$$x \in b \equiv (x \in B_\beta[X]) \,\&\, (B_\beta[X] \models \bar{\phi}[x, \gamma^*[X]])$$

$$\equiv (x \in B_\beta[X]) \,\&\, (B_\beta[X] \models \phi[x, \gamma^*[X]]) \,\&\, (x \in c^*[X])$$

$$\equiv (x \in a) \,\&\, (B_\beta[X] \models \phi[x, \gamma^*[X]]).$$

The last equivalence is obtained by remarking that $c^*[X] = a$, $a \in B_\beta[X]$ and hence $a \subseteq B_\beta[X]$. Theorem 6.2 is thus proved.

THEOREM 6.4

If X *is reducible to* M *then the axiom of replacement is valid in* $M[X]$.

Proof. Let ψ be a formula of L such that $v, w \in Fr(\psi)$ and let $g \in M[X]^{Fr(\psi)-\{v,w\}}$, $a \in M[X]$. We have to find a set b in $M[X]$ such that whenever x is in a and there is a t in $M[X]$ satisfying $M[X] \models \psi[x,t,g]$, there is a t_1 in b satisfying the same formula.

Similarly as in the previous proof we determine α, c, γ so that $c \in C_\alpha$, $\gamma \in C_\alpha^{Fr(\phi)-\{v,w\}}$ and $a = c^*[X]$, $g = \gamma^*[X]$ and put $b = B_\beta[X]$ where

$$\beta = \sup_{p\in \underline{P}} \sup_{k\in C_\alpha} \min_{\xi\in On_M} \exists d\varepsilon C_\xi \{M \models \Phi_\psi[p,k,d,\gamma]\}.$$

Note: k and d correspond here to the variables v and w respectively.

Obviously $b \in M[X]$. If x is in a then $x \in B_\alpha[X]$ and hence x has a name k in C_α. If there is a t in M[X] such that $M[X] \models \psi[x,t,g]$ then t has the form $d^*[X]$ for a suitable constant d and hence by the assumption of reducibility there is a $p \in X$ which establishes ψ at (k,d,γ) in M[X]. In view of the definition of β there is a constant d_1 in C_β such that p establishes ψ at (k,d_1,γ) in M[X]. Thus we obtain $M[X] \models \psi[k^*[X], d_1^*[X], \gamma^*[X]]$, i.e. $M[X] \models \psi[x,d_1^*[X],g]$ and we have proved that there is a $t_1 = d_1^*[X]$ in b satisfying $\psi[x,t_1,g]$ in M[X].

THEOREM 6.5

If X *is reducible to* M *then the axiom of power-sets is valid in* M[X].

Proof. Let $a \in M[X]$, $a = c^*[X]$ where $c \in C_\alpha$. If $x \subseteq a$ and $x \in M[X]$ then x has a name in C. The essential step in the proof is to show that there is an ordinal β such that each subset x of a which has a name (i.e. belongs to M[X]) has a name already in C_β.

For any constant d we consider the set

$$S(d) = \{\langle q,k\rangle \in P \times C_\alpha : M \models \Phi_{v\varepsilon w}[q,k,d]\}.$$

This set obviously belongs to M. More exactly S(d)

belongs to the power-set of $P \times C_\alpha$, taken in the sense of M. We shall denote this set $P_M(P \times C_\alpha)$. Furthermore let Σ be a family consisting of all pairs $\langle p,s \rangle \in P \times M$ such that there is a d in C satisfying the conditions:

(5) $$M \models \Phi_{v \subseteq w}[p,d,c],$$

(6) $$s = S(d).$$

The family Σ belongs to M because it is a definable subset of the set $P \times P_M(P \times C_\alpha)$ which is an element of M. For $(p,s) \in \Sigma$ let $\beta(p,s)$ be the minimal ordinal ξ such that (5) and (6) hold for a d in C_ξ and let $\beta = \sup\{\beta(p,s) : (p,s) \in \Sigma\}$. We claim that each subset x of a which belongs to $M[X]$ has a name in C_β.

Let us assume that d is a name of a set $x \in M[X]$ such that $x \subseteq a$. Since X is reducible to M there is a p satisfying (5). Thus the pair $(p,S(d))$ belongs to Σ. It follows that there is a d_1 in C_β such that

(7) $$M \models \Phi_{v \subseteq w}[p,d_1,c],$$

(8) $$S(d) = S(d_1).$$

We shall show

(9) $$d_1^*[X] = d^*[X].$$

First notice that from (7) we obtain $d_1^*[X] \subseteq c^*[X]$ because $p \in X$ and thus $d_1^*[X] \subseteq a$.

Now assume that $y \in d^*[X] = x$. Since $x \subseteq a \subseteq B_\alpha[X]$ we obtain $y \in B_\alpha[X]$ and y has a name, say k, in C_α. From

$k^*[X] \in d^*[X]$ we infer that there is a q in X which establishes $k \varepsilon d$ in $M[X]$. Hence $\langle q,k\rangle \in S(d)$ and, in view of (8), $\langle q,k\rangle \in S(d_1)$. Thus q establishes $k \varepsilon d_1$ in $M[X]$ and, since $q \in X$, we obtain $y \in d_1^*[X]$.

Let us assume conversely that $y \in d_1^*[X]$. Since $d_1^*[X] \subseteq a \subseteq B_\alpha[X]$ we can put similarly as above, $y = k^*[X]$ where $k \in C_\alpha$ and we obtain a q in X such that $\langle q,k\rangle \in S(d_1)$. From this and from (8) we infer $\langle q,k\rangle \in S(d)$ and hence $k^*[X] \in d^*[X]$, i.e. $y \in d^*[X]$. The formula (9) is thus proved. To finish the proof we denote by b the extension in $B_\beta[X]$ of the formula $v \subseteq w$ with respect to the assignment $\gamma(w) = a$. The set b is an element of $M[X]$ and consists of subsets of a. If $x \subseteq a$ and $x \in M[X]$ then x has a name in C_β and thus belongs to b. Hence b satisfies the formula $\forall x\,[x \in b \equiv x \subseteq a]$ in $M[X]$.

Theorems 4.10 and 6.3 - 6.5 show that if M is a model of ZF then so is $M[X]$ provided that X is reducible to M. For models of ZFC we have the following result:

THEOREM 6.6

If M *is a model of* ZFC *and* X *is reducible to* M, *then* $M[X]$ *is a model of* ZFC.

Proof. Let $a \in M[X]$. We shall exhibit a relation which well-orders a and is an element of $M[X]$. Since each element of $M[X]$ is a subset of a set of the form $B_\alpha[X]$ it is sufficient to exhibit a well-ordering of $B_\alpha[X]$.

Let $\leqslant$ be a relation which well-orders C_α ; the existence of $\leqslant$ follows from our assumption that the axiom of choice is valid in M. Let f be a function definable in M[X] such that $f(c,X) = c^*[X]$ for each c in C (see Lemma 5.5). Call c the earliest name of an element a of M[X] if $f(c,X) = x$ but $f(c',X) \neq x$ whenever $c' < c$. The following relation R is easily seen to be the required well-ordering of $B_\alpha[X]$: xRy if and only if the earliest name of x precedes (in the sense of the relation $\leqslant$) the earliest name of y. The proof that $R \in M[X]$ is immediate.

7. *Heuristic explanation of the construction of reducible filters*

We shall construct a reducible filter $X \subseteq P$ by considering a theory $\mathcal{T}$ formulated in RL. The theory $\mathcal{T}$ describes M[X] in the sense that for each filter F of P the set M[F] is a model of $\mathcal{T}$. A particular model of $\mathcal{T}$ can be constructed by the well-known method due to Henkin. When applying this method we consider a sequence $\{\phi_n\}$ of all sentences of RL and build a complete extension $\mathcal{T}'$ of $\mathcal{T}$ by successive steps: $\mathcal{T}' = \mathcal{T} \cup \mathcal{T}_1 \cup \mathcal{T}_2 \cup \ldots$. In the n-th step we decide of the n-th sentence ϕ_n of RL whether ϕ_n or $\neg\phi_n$ will be included in $\mathcal{T}'$. Also several other sentences have to be included in the n-th step. The inductive definition of $\mathcal{T}_n$ is carried out along with an inductive definition of a decreasing sequence $p_0 \geqslant p_1 \geqslant \ldots$ of conditions, $\mathcal{T}_n$ being the set of sentences which are true in models M[Y] for almost all Y in $[p_n]$. The words "almost all" mean here "all up to a set of first category".

As is always the case with models built by Henkin's method the complete set $\mathcal{T}'$ determines a model M[X] such that for each sentence ϕ of RL the sentence ϕ is true in M[X] if and only if

$\phi \in \mathcal{F}'$. It follows easily that all the conditions p_n belong to X, because the sentence $\underline{p}_n \varepsilon \sigma$ is true in all models $M[Y]$ where $Y \in [p_n]$. It turns out that the conditions p_n generate a maximal filter (see Lemma 9.5 below). Hence X is a maximal filter. We show that it is reducible to M. The reason for this is the following: the formula $M[X] \models \phi[\gamma^*[X]]$ is equivalent to $\phi(\gamma) \in \mathcal{F}_n$ for some n and this in turn is equivalent to

(1) $M[Y] \models \phi(\gamma)$ for almost all Y in $[p_n]$.

We shall show that this relation between ϕ, γ and p_n is definable in M, i.e. has the form $M \models \Phi_\phi[p_n,\gamma]$. The proof of definability of (1) is the most important step in the proof.

In this way we not only prove the reducibility of X to M but establish the meaning of the formulae Φ_ϕ. The formula $M \models \Phi_\phi[p,\gamma]$ says that the set of maximal filters Y in [p] for which $M[Y] \models \phi[\gamma^*[Y]]$ is co-meager (in the space $\mathcal{X}$ of all maximal filters of P). This relation is called the forcing relation.

We proceed now to the details of the proof.

8. The theory $\mathcal{T}$

The following sentences of RL are called axioms of $\mathcal{T}$:

(1) $\underline{m} \varepsilon \underline{m}'$ for $m, m' \in M$ such that $m \in m'$,

(2) $\neg(\underline{m} \varepsilon \underline{m}')$ for $m, m' \in M$ such that $m \notin m'$,

(3) $V_\alpha c$ for c in C, $\rho(c) \leq \alpha$,

(4) $c' \varepsilon c_{\alpha,\phi,\gamma} \equiv \phi^{(\alpha)}(c',\gamma)$ if ϕ is a formula of L, $v \in Fr(\phi)$, $\gamma \in C_\alpha^{Fr(\phi)-\{v\}}$, $c' \in C_\alpha$,

(5) $\forall v_0 \, \forall v_1 \, \{(\forall v_2[v_2 \, \varepsilon \, v_0 \equiv v_2 \, \varepsilon \, v_1] \equiv (v_0 \approx v_1)\}$,

(6) $\forall v_0 \, \forall v_1 \, \forall v_2 \, \{v_0 \approx v_1 \rightarrow [v_0 \, \varepsilon \, v_2 \equiv v_1 \, \varepsilon \, v_2]\}$,

(7) $\forall v \, \neg(v \, \varepsilon \, v)$,

(8) $\forall v_0 \, \forall v_1 \, \{v_0 \approx v_1 \rightarrow [V_\alpha v_0 \equiv V_\alpha v_1]\}$ for $\alpha \in On_M$,

(9) $\forall v_0[v_0 \, \varepsilon \, \sigma \rightarrow v_0 \, \varepsilon \, \underline{P}]$,

(10) $\forall v_0 \, \forall v_1[(v_0 \leq v_1 \, \& \, v_0 \, \varepsilon \, \sigma) \rightarrow v_1 \, \varepsilon \, \sigma]$,

(11) $\forall v_0 \, \forall v_1\{(v_0 \, \varepsilon \, \sigma \, \& \, v_1 \, \varepsilon \, \sigma) \rightarrow \exists v_2[v_2 \leq v_0 \, \& \, v_2 \leq v_1 \, \& \, v_2 \, \varepsilon \, \sigma]\}$,

(12) $\exists v_0(v_0 \, \varepsilon \, \sigma)$.

In the axioms (10) and (11) we used the abbreviation $v_0 \leq v_1$ for a formula of RL which expresses the fact that the ordered pair $\langle v_0, v_1 \rangle$ is an element of the set $\leq$. We should remember here that the ordering $\leq$ of P is a set of ordered pairs of which we assumed that it belongs to M. The explicit definition of $v_0 \leq v_1$ is as follows: let $\Pi'(v_m, v_n, v_p)$ be the formula "v_m is the pair v_n, v_p" i.e.

$$\Pi'(v_m,v_n,v_p) \; : \; \forall v_q[v_q \, \varepsilon \, v_m \equiv (v_q \approx v_n \vee v_q \approx v_p)]$$

where q is any integer different from m, n, p, e.g. $= m+n+p+1$. Let $\Pi(v_m,v_n,v_p)$ be the formula: v_m is the ordered pair $\langle v_n, v_p \rangle$ i.e.

$$\Pi(v_m,v_n,v_p) \; : \; \exists v_q \, \exists v_r \, [\Pi'(v_m,v_q,v_r) \, \& \, \Pi'(v_q,v_n,v_r) \, \& \, \Pi'(v_r,v_n,v_p)]$$

where $q = m + n + p + 1$, $r = m + n + p + 2$. Finally $v_0 \leq v_1$ is the formula

$$\exists v_2 \, [\Pi(v_2,v_0,v_1) \, \& \, (v_2 \, \varepsilon \, \underline{\leq})].$$

LEMMA 8.1

If X *is a filter in* P *then all the axioms* (1) - (12) *are valid in* M[X].

Proof: obvious.

It is not true that all the models of the axioms have the form M[X]. This follows for instance from the upward Skolem-Löwenheim theorem according to which there are non-denumerable models of axioms (1) - (12) whereas models M[X] are denumerable for any $X \subseteq P$.

Let us say that a pair F, G of sets of sentences of RL satisfies the postulates if

(i) $F \subseteq G$,

(ii) $\exists v_n \, \phi \in F$ where $v_n \in Fr(\phi)$ implies $\phi(c) \in G$ for some $c \in C$,

(iii) $\exists v_n \, [V_\alpha v_n \,\&\, \phi] \in F$ where $v_n \in Fr(\phi)$ implies $\phi(c) \in G$ for some $c \in C_\alpha$,

(iv) $V_\alpha c \in F$ implies $c \approx c' \in G$ for some $c' \in C_\alpha$,

(v) $c \,\varepsilon\, c' \in F$ where $c' \in C_\alpha$ implies $c \approx c'' \in G$ for some $c'' \in \cup\{C_\beta : \beta < \alpha\}$,

(vi) $c' \,\varepsilon\, \underline{m} \in F$ implies $c' \approx \underline{n} \in G$ for some $n \in m$.

A set F of sentences of RL is closed if it is consistent, complete, contains the axioms (1) - (12) and the pair F, F satisfies the postulates.

LEMMA 8.2

The theory of M[X] *(i.e. the set of sentences of* RL *which are true in* M[X]*) is closed.*

Proof: obvious.

We shall establish some properties of closed sets:

1. If ϕ is logically valid then $\phi \in F$.

Otherwise $\neg\phi$ would be in F; but each set containing $\neg\phi$ is inconsistent.

2. If ϕ, $\phi \to \psi$ are in F then so is ψ.

Otherwise $\neg\psi$ would be in F; but each set containing the sentences ϕ, $\phi \to \psi$, $\neg\psi$ is inconsistent.

3. If ϕ is logically equivalent to ψ then $\phi \in F$ if and only if $\psi \in F$.

This follows from 1 and 2.

4. If v is the unique free variable of ϕ then $\forall v\, \phi \in F$ is equivalent to: $\phi(c) \in F$ for each c in C.

Since $\forall v\, \phi \to \phi(c)$ is logically true we obtain the implication $\to$ from 1 and 2. If $\forall v\, \phi \notin F$ then $\exists v\, \neg\phi \in F$ hence $\neg\phi(c) \in F$ for some c and so it is not true that $\phi(c) \in F$ for each c.

5. If v is the unique free variable of ϕ then $\exists v\, \phi \in F$ if and only if there is a c in C such that $\phi(c) \in F$.

Proof: similar to 4.

6. If ϕ is a formula, $\gamma \in C^{Fr(\phi)-\{v\}}$, then $c_1 \approx c_2 \in F$ implies $\phi(c_1,\gamma) \equiv \phi(c_2,\gamma) \in F$ for each ϕ.

Proof. If $\phi(\gamma)$ is the atomic formula $v \varepsilon c$ or $c \varepsilon v$ then 6. follows from axioms (5), (6) and the above remarks 1 - 4.

If ϕ is the formula $v \varepsilon \dot{v}$, 6. follows from axiom (7) : $\neg(c_1 \varepsilon c_1) \in F$ hence $c_1 \varepsilon c_1 \to \psi \in F$ for any ψ and we obtain $c_1 \varepsilon c_1 \to c_2 \varepsilon c_2 \in F$. Similarly $c_2 \varepsilon c_2 \to c_1 \varepsilon c_1 \in F$.

If $\phi(\gamma)$ is the formula $v \approx c$ or $c \approx v$ we prove 6. using axiom (6). If ϕ is the formula $v \approx v$, we again use (6). If ϕ is the formula $V_\alpha v$ we use axiom (8).

Let us now assume 6. for two formulae ϕ', ϕ''. Using tautologous formulae of propositional logic we immediately obtain 6. for the formulae $\neg\phi'$ and $\phi' \& \phi''$. Also using 4. we obtain that $c_1 \approx c_2 \in F$ implies $\forall w\, [\phi'(c_1,\bar{\gamma}) \equiv \phi'(c_2,\bar{\gamma})] \in F$ where $\mathrm{Dom}(\bar{\gamma}) = \mathrm{Fr}(\phi') - \{v,w\}$. Using the fact that the taulogous sentence $\forall w\, (\phi_1 \equiv \phi_2) \to [\forall w\, \phi_1 \equiv \forall w\, \phi_2]$ is in F we obtain that $c_1 \approx c_2 \in F$ implies $[\forall w\ \ \phi'(c_1,\bar{\gamma})] \equiv [\forall w\ \phi'(c_2,\bar{\gamma})] \in F$.

7. If m_1, $m_2 \in M$, $m = \{m_1,m_2\}$ then $\Pi'(\underline{m},\underline{m}_1,\underline{m}_2) \in F$.

Proof. Using the definition of Π' and 4. above we see that we have to prove for each c in C the following two formulae:

(*) $c \varepsilon \underline{m} \to (c \approx \underline{m}_1 \vee c \approx \underline{m}_2) \in F$,

(**) $(c \approx \underline{m}_1 \vee c \approx \underline{m}_2) \to c \varepsilon \underline{m} \in F$.

To prove the second formula we observe that $\underline{m}_i \varepsilon \underline{m} \in F$, $i = 1,2$, by (1) and hence $c \approx \underline{m}_i \to c \varepsilon \underline{m} \in F$ for $i = 1, 2$. From this we obtain (**) by a propositional tautology. The formula (*) is shown by contradiction. If (*) were false, then, by completeness, the three sentences $c \varepsilon \underline{m}$, $\neg(c \approx \underline{m}_1)$, $\neg(c \approx \underline{m}_2)$ would belong to F. Using postulate (vi) we would obtain from $c \varepsilon \underline{m} \in F$ that $c \approx \underline{m}_1 \in F$ or $c \approx \underline{m}_2 \in F$ and F would be inconsistent.

8. If $m_1, m_2 \in M$, $m = \langle m_1, m_2 \rangle$ then

$$\forall v\, [\Pi(v, \underline{m}_1, \underline{m}_2) \equiv (v \approx \underline{m})] \in F.$$

Proof is similar to 7.

LEMMA 8.3

If F *is closed then the set* $X = \{p \in \underline{P} : \underline{p} \,\varepsilon\, \sigma \in F\}$ *is a filter in* P.

Proof. $X \neq \emptyset$ because by axiom (12) and 5. there is a constant c such that $c \,\varepsilon\, \sigma \in F$ whence, by axiom (9) and properties 2, 4 above, $c \,\varepsilon\, \underline{P} \in F$ whence by postulate (vi) $c \approx \underline{p} \in F$ for some p in P. Thus, by 6, $\underline{p} \,\varepsilon\, \sigma \in F$ and so $p \in X$.

Next, let $p, q \in X$. Hence $\underline{p} \,\varepsilon\, \sigma \in F$ and $\underline{q} \,\varepsilon\, \sigma \in F$. From axiom (11) we infer that the existential sentence

$$\exists v_2\, (v_2 \leq \underline{p} \,\&\, v_2 \leq \underline{q} \,\&\, v_2 \in \sigma)$$

is in F. Hence there is a constant c such that the sentences

$$c \leq \underline{p}, \quad c \leq \underline{q}, \quad c \,\varepsilon\, \sigma$$

belong to F. From $c \,\varepsilon\, \sigma \in F$ and axiom (9) we infer $c \,\varepsilon\, \underline{P} \in F$ and hence, by postulate (vi) $c \approx \underline{r} \in F$ for some r in P. Now we obtain $\underline{r} \,\varepsilon\, \sigma \in F$, hence $r \in X$ and $\underline{r} \leq \underline{p} \in F$ and $\underline{r} \leq \underline{q} \in F$. These formulae prove that if $s = \langle r,p \rangle$, $t = \langle r,q \rangle$ then $\underline{s} \,\varepsilon\, \underline{\leq} \in F$ and $\underline{t} \,\varepsilon\, \underline{\leq} \in F$ whence $s, t \in \leqslant$. Thus $r \leqslant p$ and $r \leqslant q$.

Finally let $p \in X$ and $p \leqslant q$. We infer similarly as above that $\underline{p} \leq \underline{q} \in F$ whence, in view of $\underline{p} \,\varepsilon\, \sigma \in X$ and axiom (10) $\underline{q} \,\varepsilon\, \sigma \in F$ and so $q \in X$.

LEMMA 8.4

If F *is a closed set of sentences and* $X = \{p \in P : \underline{p} \varepsilon \sigma \in F\}$ *then* M[X] *is a model of* F.

Proof. First we construct a relational system A in which all sentences $\phi \in F$ are true and then prove that after dividing A by a congruence we obtain a relational system isomorphic with M[X]. We obtain A by a standard method, due to Henkin, of constructing relational systems from constants.

The universe of A will be C; for each c in C the interpretation of c will be c itself. The binary predicates $\approx$, ε will be interpreted as the relations I, E defined as follows:

(i) $c_1 I c_2 \equiv (c_1 \approx c_2 \in F) \quad c_1 E c_2 \equiv (c_1 \varepsilon c_2 \in F)$.

Finally the unary predicates V_α will be interpreted as sets

(ii) $$A_\alpha = \{c \in C : V_\alpha c \in F\}.$$

It follows from remark 6 above that I is a congruence in A.

We prove by induction that if ϕ is a formula of RL and $\gamma \in C^{Fr(\phi)}$ then

(iii) $$A \models \phi[\gamma] \equiv \phi(\gamma) \in F.$$

For atomic formulae ϕ this follows directly from the definition. If the equivalence is true for ϕ and ϕ_1 then using completeness of F and 4 we immediately infer that it is also valid for $\neg\phi$, $\phi \,\&\, \phi_1$ and $\forall v\, \phi$ where v is any variable.

Thus A and also A divided by the equivalence I which we shall denote by A/I are relational systems in which all sentences of F are true.

We prove the following statement: for each limited formula ϕ and each $\gamma \in C^{Fr(\phi)}$

(iv) $$M[X] \models \phi[\gamma^*[X]] \equiv A \models \phi[\gamma].$$

Let us write $(\psi,\delta) < (\phi,\gamma)$ if ϕ, ψ are limited formulae, γ, δ sequences of constants and $\psi(\delta) \prec \phi(\gamma)$ where $\prec$ is the relation defined in Section 5. It will be sufficient to prove (iv) under the assumption that it is valid for all pairs $(\psi,\delta) < (\phi,\gamma)$.

If $\phi(\gamma)$ is minimal with respect to the relation $\prec$ then $\phi(\gamma)$ is one of the sentences $\underline{0} \in \underline{0}$, $\underline{0} \approx \underline{0}$, $V_0\underline{0}$ and the truth of (iv) is easy to verify.

Let us assume that $\phi(\gamma)$ is not minimal. If ϕ contains logical connectives then either (case 1) $\phi(\gamma) = \neg\phi_1(\gamma)$ or (case 2) $\phi(\gamma) = \phi_1(\gamma \mid Fr(\phi_1)) \;\&\; \phi_2(\gamma \mid Fr(\phi_2))$ or (case 3) $\phi(\gamma) = \forall v_j\,[V_\alpha v_j \to \phi_1(\gamma)]$ and (iv) is true for formulae ϕ_1, ϕ_2 and arbitrary sequences γ_1, γ_2 of constants such that $(\phi_1, \gamma_1) < (\phi,\gamma)$ and $(\phi_2, \gamma_2) < (\phi,\gamma)$. In case 1 $(\phi_1,\gamma) < (\phi,\gamma)$ hence (iv) is valid for the pair (ϕ_1,γ) and taking negations on both sides we obtain (iv) for the pair (ϕ,γ). In case 2 $(\phi_i, \gamma \mid Fr(\phi_i)) < (\phi,\gamma)$ for $i = 1,2$ and so (iv) is valid for the pairs $(\phi_i, \gamma \mid Fr(\phi_i))$, $i = 1,2$. Taking conjunctions on both sides of the resulting equivalences we obtain (iv) for the formula ϕ. In case 3 the left-hand side of (iv) is equivalent to the statement: for each x in $B_\alpha[x]$

$$M[X] \models \phi_1[x,\gamma^*[X]].$$

Since each element of $B_\alpha[X]$ has a name in C_α and since $c^*[X] \in B_\alpha[X]$ for $c \in C_\alpha$ we infer, using the inductive assumption, that this statement is equivalent to: for each c in C_α, $A \models \phi_1 [c,\gamma]$. Now let c' be an arbitrary constant. If $A \not\models V_\alpha[c']$ then c' satisfies $V_\alpha v_j \to \phi_1(\gamma)$ in A. Otherwise by postulate (iv) there is a constant c in C_α such that $c \, I \, c'$ and since, as we proved above, $A \models \phi_1[c,\gamma]$, we obtain $A \models \phi_1[c',\gamma]$. Hence each c' satisfies in A the formula $V_\alpha v_j \to \phi_1(\gamma)$ and we obtain the right-hand side of (iv).

The converse implication is proved similarly.

It remains to prove (iv) in the case when ϕ is an atomic formula. We have several cases to consider.

Case 1. $\phi(\gamma)$ is the sentence $c_1 \varepsilon c_2$. Subcase (a) : $\rho(c_1) \geq \rho(c_2)$. In this case the left-hand side of (iv) is equivalent to $c_1^*[X] \in c_2^*[X]$. Let c_3 be a constant of a possibly smaller order such that $c_1^*[X] = c_3^*[X]$. Thus $\rho(c_3) < \rho(c_2)$ and it follows that the sentences $c_1 \approx c_3$ and $c_3 \varepsilon c_2$ precede the sentence $c_1 \varepsilon c_2$ in the ordering $\prec$.

From the inductive assumption we obtain therefore $A \models c_1 \approx c_3$ and $A \models c_3 \varepsilon c_2$ whence $c_1 \approx c_3 \in F$, $c_3 \varepsilon c_2 \in F$ and therefore $c_1 \varepsilon c_2 \in F$ i.e., $A \models c_1 \varepsilon c_2$. The implication can obviously be reversed.

Subcase (b) : $\rho(c_1) < \rho(c_2)$. We denote $\rho(c_2)$ by $\alpha + 1$ and discuss separately the possible forms of c_2:

Sub-subcase (b1) : $c_2 = \underline{m}$ where $m \in M$ and $rk(m) < \alpha$. In this case the left-hand side of (iv) is equivalent to $c_1^*[X] = n$

where $n \in m$. Putting $c_3 = \underline{n}$ we have $c_1 \approx c_3 \prec c_1 \varepsilon c_2$ and so, by inductive assumption, $A \models c_1 \approx \underline{n}$ whence $c_1 \approx \underline{n} \in F$ and so $c_1 \varepsilon \underline{m} \in F$, i.e., $A \models c_1 \varepsilon c_2$.

Conversely, by postulate (vi), the formula $A \models c_1 \varepsilon c_2$ implies $c_1 \approx \underline{n} \in F$ for some n in m and the implications above can be reversed.

Sub-subcase (b2) : $c_2 = \sigma$. Here the left-hand side of (iv) is equivalent to $c_1^*[X] = p$ where $p \in X$. Since $\rho(\underline{p}) < \rho(\sigma)$ we can repeat the previous proof. Similarly the right-hand side of (iv) is equivalent to $c_1 \varepsilon \sigma \in F$ from which we infer, using axiom (9) and postulate (vi) that $c_1 \approx \underline{p} \in F$ for some p in X. Hence we obtain $c_1^*[X] = p$ because $c_1 \approx \underline{p} \prec c_1 \varepsilon \sigma$.

Sub-subcase (b3) : $c_2 = c_{\alpha,\phi,\gamma}$ where ϕ is a formula of L, $v \in Fr(\phi)$ and $\gamma \in C_\alpha^{Fr(\phi)-\{v\}}$. In this case the left-hand side of (iv) is equivalent to $c_1^*[X] \in E_{\phi,\gamma^*[X],B_\alpha[X]}$ i.e. to $B_\alpha[X] \models \phi[c_1^*[X], \gamma^*[X]]$. We can replace here ϕ by $\phi^{(\alpha)}$ and $B_\alpha[X]$ by $M[X]$ because the satisfaction of a formula in $B_\alpha[X]$ is equivalent to the satisfaction of the relativized formula $\phi^{(\alpha)}$ in $M[X]$. Now $\phi^{(\alpha)}(c_1,\gamma) \prec c_1 \varepsilon c_2$ because orders of the constants occurring in $\phi^{(\alpha)}(c_1,\gamma)$ are $\leq \alpha$ and all the unary predicates which occur in $\phi^{(\alpha)}$ have indices α whereas $\rho(c_1 \varepsilon c_2) = \rho(c_2) = \alpha + 1$. Hence we can use the inductive assumption and obtain $A \models \phi^{(\alpha)}[c_1,\gamma]$, i.e., $\phi^{(\alpha)}(c_1,\gamma) \in F$. Using axiom (4) we obtain $c_1 \varepsilon c_{\alpha,\phi,\gamma} \in F$ which is the same as $c_1 \varepsilon c_2 \in F$.

All these steps can obviously be reversed.

Formula (iv) is thus proved in case 1.

Case 2. $\phi(\gamma)$ is the formula $c_1 \approx c_2$. We can assume that $\rho(c_1) \leq \rho(c_2) = \alpha + 1$. The left-hand side of (iv) is equivalent to $c_1^*[X] = c_2^*[X]$ i.e., to $\forall x_{B_\alpha[X]} (x \in c_1^*[X] \equiv x \in c_2^*[X])$ which in turn is equivalent to $\forall c_{C_\alpha} (c^*[X] \in c_1^*[X] \equiv c^*[X] \in c_2^*[X])$. Now we notice that if $c \in C_\alpha$ and $\rho(c_1) \leq \alpha$ then $\rho(c \varepsilon c_1) < \alpha + 1 = \rho(c_1 \approx c_2)$; if $\rho(c_1) = \alpha + 1$ then the formulae $c \varepsilon c_1$ and $c_1 \approx c_2$ have the same orders but $c \varepsilon c_1 \prec c_1 \approx c_2$ according to the definition of $\prec$. Thus in both cases $c \varepsilon c_1 \prec c_1 \approx c_2$. Similarly $c \varepsilon c_2 \prec c_1 \approx c_2$. It follows now by the inductive assumption that the left-hand side of (iv) is equivalent to $\forall c_{C\alpha} A \models (c \varepsilon c_1 \equiv c \varepsilon c_2)$. From axiom (5) we see that $A \models c_1 \approx c_2$ implies the formula $A \models (c \varepsilon c_1 \equiv c \varepsilon c_2)$. Hence the right-hand side of (iv) implies the left-hand side.

It remains to prove that if $A \not\models c_1 \approx c_2$ then there is a c in C_α such that $A \not\models c \varepsilon c_1 \equiv c \varepsilon c_2$.

Let us assume $A \not\models c_1 \approx c_2$, i.e., $\neg(c_1 \approx c_2) \in F$. Using axiom (5) and property 5 of closed sets we obtain a constant c' such that either $c' \varepsilon c_1 \in F$ and $\neg(c' \varepsilon c_2) \in F$ or $\neg(c' \varepsilon c_1) \in F$ and $c' \varepsilon c_2 \in F$. We can limit ourselves to the first case only. We use postulate (v) and infer from $c' \varepsilon c_1 \in F$ that $c' \approx c \in F$ and $c \varepsilon c_1 \in F$ for some $c \in C_\alpha$. Hence $c \varepsilon c_1 \in F$ and $\neg(c \varepsilon c_2) \in F$ and therefore $A \not\models (c \varepsilon c_1 \equiv c \varepsilon c_2)$.

Case 3. $\phi(\gamma)$ is the formula $V_\alpha c$. The left-hand side of (iv) is equivalent to $c^*[X] \in B_\alpha[X]$, i.e., to $c^*[X] = c_1^*[X]$ for some c_1 in C_α. Since $c \approx c_1 \prec V_\alpha c$, according to the definition of $\prec$, we obtain $A \models c \approx c_1$, i.e., $c \approx c_1 \in F$. By axiom (3) $V_\alpha c_1 \in F$ whence $V_\alpha c \in F$ and therefore $A \models V_\alpha c$. Conversely, if $A \models V_\alpha c$, then, by postulate (iv), $c \approx c_1 \in F$ for some c_1 in C_α and the previous steps can be reversed.

The proof of (iv) is thus complete.

In order to finish the proof of Lemma 8.4 we remark that (iv) implies the equivalences

$$M[X] \models c_1^*[X] \in c_2^*[X] \equiv A \models c_1 \; E \; c_2,$$

$$M[X] \models c_1^*[X] = c_2^*[X] \equiv A \models c_1 \; I \; c_2,$$

$$M[X] \models c^*[X] \in B_\alpha[X] \equiv A \models A_\alpha(c).$$

These equivalences show that $M[X]$ is isomorphic to A/I.

Lemma 8.4 shows that we can obtain filters by constructing closed sets of sentences. In the next section we shall construct such a set and then show that the resulting filter X is reducible to M.

9. *Construction of a closed set* F.

We consider a partially ordered set P as described in Section 2 and denote by $\mathcal{X}$ the space of its maximal filters.

Let K be a σ-additive field of subsets of $\mathcal{X}$ and I a σ-additive ideal in K. For each sentence ϕ of RL we put

$$F_\phi = \{X \in \mathcal{X} : M[X] \models \phi\}.$$

We shall assume that K and I have the following properties:

(A) $F_\phi \in K$ for each sentence ϕ of RL,

(B) If $p \in P$ then $[p] \notin I$,

(C) If $H \in K - I$ then there is a p in P such that $[p] - H \in I$,

(D) For each formula ϕ of RL the binary relation $[p]-F_{\phi(\gamma)} \in I$, where $p \in P$, and $\gamma \in C^{Fr(\phi)}$, is definable in M.

We shall show in Sections 10-12 that (A) - (D) are satisfied if K is the field of Borel sets in $\mathcal{X}$ and I the ideal of meager sets.

The binary relation from condition (D) will be written as $p \Vdash \phi(\gamma)$ and read "p forces $\phi(\gamma)$".

We note some simple consequences of the definitions and assumptions (A) - (D).

LEMMA 9.1

If $p,q \in P$ *and* $p \nleq q$, *then* $[p] - [q] \notin I$.

Proof. There is r in P such that $r \leq p$ and r and q are incompatible, hence $[r] \subseteq [p] - [q]$ and thus if $[p] - [q]$ were in I we would have a contradiction with (B).

LEMMA 9.2

$$F_{\neg\phi} = \mathcal{X} - F_{\phi} \; ; \qquad F_{\phi \& \psi} = F_{\phi} \cap F_{\psi} \; ;$$

$$F_{\forall x\ \phi} = \bigcap_{c \in C} F_{\phi(c)} \; , \quad F_{\forall_{\alpha} x\ \phi} = \bigcap_{c \in C_{\alpha}} F_{\phi(c)} .$$

Proof results immediately from the definition of F_{ϕ}.

Let us arrange in an infinite sequence $\{\phi_n\}_{n\in\omega}$ all sentences of RL. Let $\{\psi_n\}_{n\in\omega}$ be a sequence consisting of all axioms (1) - (12). For any finite set S of sentences we denote by $\wedge S$

their conjunction and by $\{S,\alpha,\beta,\ldots,\gamma\}$ the set $S \cup \{\alpha,\beta,\ldots,\gamma\}$. In order to construct a closed set we try to define an increasing sequence $\{F_n\}_{n\in\omega}$ of finite sets of sentences such that the following requirements be met for each $n > 0$:

(R_1) F_n is consistent,

(R_2) $\psi_{n-1} \in F_n$,

(R_3) either ϕ_{n-1} or $\neg\phi_{n-1}$ belongs to F_n,

(R_4) the pair (F_{n-1},F_n) satisfies the postulates.

It is clear that if these requirements are met, the union $\cup F_n$ will be closed.

LEMMA 9.3

There are infinite sequences $\{p_n\}_{n\in\omega},\{F_n\}_{n\in\omega}$ *consisting of conditions and finite sets of sentences respectively such that, for each integer* n, $p_n \Vdash \wedge F_n$ *and* F_n *satisfies the requirements* (R_1) - (R_4).

Proof. For $n = 0$ we take $F_0 = \emptyset$ and define p_0 to be any element of P. Let us assume that p_n and F_n are already constructed. We construct F_{n+1} by adjoining to F_n several sentences. First of all we adjoin ψ_n. Since ψ_n is true in $M[X]$ for each X, the formula $p_n \models \wedge\{F_n,\psi_n\}$ continues to hold.

Next we try to adjoin ϕ_n or $\neg\phi_n$ to $\{F_n,\psi_n\}$. For each X in $\mathcal{X}$ we either have $M[X] \models \phi_n$ or $M[X] \models \neg\phi_n$. Thus the set $[p_n]$ decomposes into two parts consisting of maximal filters X for which the former or the latter formula holds. Both of these parts belong to K but it cannot be the case that both of them are in I

since otherwise $[p_n]$ itself would belong to I. Hence one of these parts, call it H, is in $K - I$ and hence by (C) there is a $p_n' < p_n$ such that $[p_n'] - H \in I$. We denote by F_n' the set $\{F_n, \psi_n, \phi_n\}$ if ϕ_n is true in the models $M[X]$ where $X \in H$, and the set $\{F_n, \psi_n, \neg\phi_n\}$ if $\neg\phi_n$ is true in these models. Thus we obtain

$$p_n' \Vdash \wedge F_n'$$

and F_n' satisfies the requirements $(R_1) - (R_3)$.

In order to satisfy the requirement (R_4) we have still to add various sentences to F_n' and restrict, if necessary, the condition p_n'.

Let us first add new sentences to F_n' so as to obtain a set which together with F_n satisfies the postulate (ii). To achieve this we enumerate the existential sentences (i.e. sentences beginning with the symbols $\neg\forall v_j$ which belong to F_n. Let these sentences be

(*) $$\exists w\theta, \exists w'\theta', \ldots, \exists w_k\theta^{(k)}.$$

For each X in $[p_n']$ the sentence $\exists w\theta$ is true in $M[X]$ and so for each X there exists a constant c_X such that $M[X] \models \theta(c_X)$. The set $[p_n']$ is thus decomposed into a denumerable union of sets $S_c = \{X \in [p_n'] : M[X] \models \theta(c)\}$. By (A) these sets belong to K but it cannot be the case that they are all elements of I. Hence there is a constant c such that $S_c \notin I$ and hence by (C) there is a condition $p_n'' < p_n'$ such that $[p_n''] - S_c \in I$. It follows that $p_n'' \Vdash \theta(c)$. Thus adjoining $\theta(c)$ to F_n' we obtain a set F_n'' such that $p_n'' \Vdash \wedge F_n''$. Repeating this process again k times we finally obtain a set $F_n^{(k)}$ and a condition $p_n^{(k)}$ such that

$p_n^{(k)} \Vdash \wedge F_n^{(k)}$ and $F_n^{(k)}$ contains sentences $\theta(c)$, $\theta'(c')$,..., $\theta^{(k)}(c^{(k)})$ for each of the formulae (*). Thus the pair $(F_n, F_n^{(k)})$ satisfies postulate (ii).

The procedure for the remaining postulates is very similar. In case of postulate (iii) we consider limited existential sentences, i.e. sentences of the form $\exists_\alpha w\ \zeta$ which belong to F_n and for each such sentence find a constant c in C_α and a condition $p_n''' \leqslant p_n^{(k)}$ which forces $\zeta(c)$. In case of postulate (iv) we consider sentences of the form $V_\alpha c$ which belong to F_n and find for each such sentence a constant c' in C_α such that $c \approx c'$ can be adjoined to the sets previously constructed. In case of postulate (v) we consider sentences $c \varepsilon c'$ in F_n and find for each of them a constant c'' of order $< \rho(c')$ such that $c \approx c''$ can be adjoined. Finally in case of postulate (vi) we deal with sentences of the form $c \varepsilon \underline{m}$ which belong to F_n and find for each such sentence an element n' of m such that the sentence $c \approx \underline{n}'$ can be adjoined.

Lemma 9.3 is thus proved.

The sequences $\{p_n\}_{n\in\omega}$ and $\{F_n\}_{n\in\omega}$ constructed in Lemma 9.3 determine two filters: one is the filter X_0 generated by the conditions p_n and the other is the filter $X = \{p : \underline{p} \varepsilon \sigma \in F\}$ where $F = \cup F_n$. We shall show that these filters are identical. First we note the useful

LEMMA 9.4

If $\{p_n\}_{n\in\omega}$, $\{F_n\}_{n\in\omega}$ *are sequences satisfying Lemma 9.3 and* $F = \cup F_n$ *then* $\phi \in F \equiv \exists n\ (p_n \Vdash \phi)$ *for each sentence* ϕ *of* RL.

Proof. $\phi \in F \equiv \exists n\ (\phi \in F_n) \to \exists n\ (p_n \Vdash \phi)$ because the sentence $\wedge F_n \to \phi$ is logically true whenever $\phi \in F_n$. Conversely, if $\phi \notin F$, then $\neg\phi \in F$ and hence, by the above proof, $p_n \Vdash \neg\phi$ for some n. Assuming $p_m \Vdash \phi$ and putting $k = \max(m,n)$ we would obtain $p_k \Vdash \neg\phi$ and $p_k \Vdash \phi$ which is impossible by (B).

LEMMA 9.5

If $\{p_n\}_{n\in\omega}$ *and* $\{F_n\}_{n\in\omega}$ *are as in Lemma 9.3 and* $F = \cup F_n$ *then the filter* $X = \{p \in P : \underline{p} \in \sigma \in F\}$ *is identical with the filter generated by the sequence* $\{p_n\}$ *; moreover* X *is maximal.*

Proof. Let $p \in P$. For each Y in $[p]$ we have $p \in Y$ and so $M[Y] \models \underline{p} \in \sigma$. Hence $[p] - F_{p\varepsilon\sigma} = \emptyset$ and $p \Vdash \underline{p} \in \sigma$. For $p = p_n$ we obtain $p_n \in \sigma \in F$. We have thus shown that the filter generated by the p_n's is contained in X.

Next we show that if $p \in X$ then p belongs to the filter generated by the conditions p_n.

Since $\underline{p} \in \sigma \in F$ there is an integer n such that the formula $\wedge F_n \to (\underline{p} \in \sigma)$ is logically true and so $p_n \Vdash \underline{p} \in \sigma$. We claim that $p_n \leq p$. Otherwise there would exist a $q \leq p_n$ such that q and p are incompatible. Hence no filter Y in $[q]$ would satisfy $p \in Y$, i.e., the difference $[q] - \{Y \in \mathcal{X} : M[Y] \models \underline{p} \in \sigma\}$ would be equal to $[q]$. From $p_n \Vdash \underline{p} \in \sigma$ we obtain however $[p_n] - \{Y \in \mathcal{X} : M[Y] \models \underline{p} \in \sigma\} \in I$ and so $[q] - \{Y \in \mathcal{X} : M[Y] \models \underline{p} \in \sigma\} \in I$ because $[q] \subseteq [p_n]$. Thus we would obtain the result that $[q] \in I$ which is impossible.

Finally we show that X is maximal. Let Y be a filter in P such that $X \subset Y$ and assume that $p \in Y - X$. Hence $\neg(\underline{p} \in \sigma) \in F$ and therefore $p_n \Vdash \neg(\underline{p} \in \sigma)$ for an integer n. Since

$p \in Y$, the conditions p and p_n are compatible; let $r < p_n$ and $r < p$. Since $[r] - F_{\neg(\underline{p}\varepsilon\sigma)} \subseteq [p_n] - F_{\neg(\underline{p}\varepsilon\sigma)}$ we obtain $[r] \cap F_{\underline{p}\varepsilon\sigma} \in I$. On the other hand the relation $r < p$ proves that for each Y in $[r]$ the formula $p \in Y$ and hence also the formula $M[Y] \models \underline{p} \varepsilon \sigma$ is true. Thus $[r] - F_{\underline{p}\varepsilon\sigma} = \emptyset$, $[r] \subseteq F_{\underline{p}\varepsilon\sigma}$ which together with the previous relation shows that $[r] \in I$. Since this contradicts the assumption (B), Lemma 9.5 is proved.

Taking our lemmas together we obtain

THEOREM 9.6

There exists a sequence $\{p_n\}_{n\in\omega}$ *of conditions such that the set* $F = \{\phi : \exists n\ p_n \Vdash \phi\}$ *is closed. The filter* $X = \{p \in P : \underline{p} \varepsilon \sigma \in F\}$ *is reducible to* M *and maximal. It is identical with the filter generated by the conditions* p_n.

Proof. Let $\{p_n\}_{n\in\omega}$ and $\{F_n\}_{n\in\omega}$ be the sequences constructed in Lemma 9.3 and put $F = \cup F_n$. From 9.3 it follows that F is closed and from 9.4 that it coincides with the set $\{\phi : \exists n\ (p_n \Vdash \phi)\}$. From 8.4 it follows that $M[X] \models \phi[\gamma^*[X]] \equiv \phi(\gamma) \in F \equiv \exists n\ p_n \Vdash \phi(\gamma)$. Since each p_n belongs to X and each element p of X is $\geq p_n$ for some n we can write this condition as $\exists p \in X\ p \Vdash \phi(\gamma)$. In view of the assumption (D) there is a formula Φ_ϕ such that $p \Vdash \phi(\gamma) \equiv M \models \Phi_\phi[p,\gamma]$ and so X is reducible to M. The remaining two statements follow from 9.5.

10. Verification of assumptions (A), (B) *and* (C)

Let K be the field of Borel subsets of $\mathcal{X}$ and I the ideal of meager sets. We are going to prove that the assumptions

(A) - (D) of Section 9 are satisfied.

(B) follows from Baire category theorem (see 2.5). The proof of (C) is as follows: Each non-meager Borel set H has the form $(G - N) \cup N'$ where N, N' are meager and G is open and not empty (see Kuratowski [66], p.88). Hence if $[p] \subset G$, then $[p] - H \subseteq N$ and therefore $[p] - H$ is meager.

Proof of (A). From Lemma 9.2 it follows that if F_ϕ is Borel for each sentence ϕ containing less than n symbols for logical operations then it is true for the case when ϕ contains n such symbols. Thus it is sufficient to prove (A) for atomic sentences. It is more convenient to prove it more generally for limited sentences. We show that if ϕ is a limited sentence and for each $\psi \prec \phi$ the set F_ψ is Borel then so is F_ϕ.

The case when ϕ has no predecessors with respect to $\prec$ is trivial because ϕ is then one of the sentences $\underline{0} \in \underline{0}$, $\underline{0} \approx \underline{0}$, $V_0\underline{0}$ and F_ϕ is either the void set or the whole space $\mathcal{X}$.

Now let us assume that ϕ has predecessors. The cases when ϕ contains symbols for logical operations can be disposed of as above. Let ϕ now be atomic. We have three cases to consider.

1) $\phi = c_1 \varepsilon c_2$. We distinguish two subcases:

1a) $\rho(c_1) \geqslant \rho(c_2)$,

1b) $\rho(c_1) < \rho(c_2)$

Subcase 1a). Put $\rho(c_2) = \alpha$. By definition $X \in F_\phi \equiv c_1{*}[X] \in c_2{*}[X] \equiv \exists c \in C_\alpha[(c_1{*}[X] = c{*}[X]) \,\&\, (c{*}[X] \in c_2{*}[X])]$ whence

$$F_\phi = \bigcup_{c \in C_\alpha} (F_{c \varepsilon c_2} \cap F_{c \approx c_1}).$$

Since $c \varepsilon c_2$ and $c \approx c_1$ precede $c_1 \varepsilon c_2$ in the ordering $\prec$ the result follows by inductive assumption.

Subcase 1b). If $c_2 = \underline{m}$ then we show similarly that $F_\phi = \cup\{F_{c_1 \approx \underline{n}} : n \in m\}$ whence the result follows because $c_1 \approx \underline{n} \prec c_1 \varepsilon c_2$.

If $c_2 = \sigma$ then $X \in F_\phi$ is equivalent to $\exists p \in P\,[X \in [p]\ \&\ (c_1^*[X] = p)]$. Hence $F_\phi = \bigcup_{p \in P} ([p] \cap F_{c_1 \approx \underline{p}})$ and F_ϕ is Borel because $c_1 \approx \underline{p} \prec c_1 \varepsilon \sigma$.

If $c_2 = c_{\alpha,\phi,\gamma}$ then $X \in F_\phi$ is equivalent to $X \in F_{\phi^{(\alpha)}(c_1,\gamma)}$ and again the inductive assumption is applicable because $\phi^{(\alpha)}(c_1,\gamma) \prec c_1 \varepsilon c_2$.

Case 2. $\phi = c_1 \approx c_2$. We can assume that $\rho(c_1) \leq \rho(c_2) = \alpha$. The relation $X \in F_\phi$ is equivalent to

$$X \in \bigcap_{c \in C_\alpha} [F_{c \varepsilon c_1} \cap F_{c \varepsilon c_2}] \cup [(\mathfrak{X} - F_{c \varepsilon c_1}) \cap (\mathfrak{X} - F_{c \varepsilon c_2})]$$

whence we reduce the theorem to the case 1.

Case 3. $\phi = V_\alpha c$. Put $\rho(c) = \alpha$. In this case $X \in F_\phi$ is equivalent to $c^*[X] = c_1^*[X]$ for some $c_1 \in C_\alpha$ and hence $F_\phi = \bigcup_{c_1 \in C_\alpha} F_{c \approx c_1}$ whence the theorem is reduced to Case 2. Assumption (A) is thus verified.

Before verifying assumption (D) we must establish some properties of the forcing relation.

11. *Properties of the forcing relation*

We denote by ϕ, ψ sentences of RL and by p, q, r elements of P. By $\phi(v,w,\ldots)$ we denote formulae of RL all of whose free variables are among $v, w, \ldots$.

LEMMA 11.1

If $p \leqslant q$ *and* $q \Vdash \phi$ *then* $p \Vdash \phi$.

Proof. $[p] - F_\phi \subseteq [q] - F_\phi$, hence if the right-hand side belongs to I then so does the left.

LEMMA 11.2

If $\phi \rightarrow \psi$ *is logically valid then* $p \Vdash \phi$ *implies* $p \Vdash \psi$.

Proof. $F_\phi \subseteq F_\psi$ and hence $[p] - F_\psi \subseteq [p] - F_\phi$.

LEMMA 11.3

If ϕ *and* ψ *are logically equivalent then* $p \Vdash \phi$ *is equivalent to* $p \Vdash \psi$.

This follows from 11.2.

LEMMA 11.4

$p \Vdash \phi \,\&\, \psi$ *is equivalent to* $(p \Vdash \phi) \,\&\, (p \Vdash \psi)$.

Proof. $[p] - F_{\phi\&\psi} = [([p] - F_\phi) \cup ([p] - F_\psi)]$; it is now sufficient to note that the union of two sets belongs to I if and only if each of these sets does.

LEMMA 11.5

$p \Vdash \neg\phi$ *is equivalent to* $\forall q_{\leqslant p}\ (q \nVdash \phi)$.

Proof. The left-hand side is equivalent to $[p] \cap F_\phi \in I$; hence if the left-hand side is true and $q \leqslant p$ then $[q] \cap F_\phi \in I$

and so $[q] - F_\phi \notin I$ by (B). If the left-hand side is false then by (C) there is q such that $[q] - ([p] \cap F_\phi) \in I$ and we obtain $([q] - [p] \cup ([q] - F_\phi) \in I$. Since by (C) $[q] - [p] \in I$ if and only if $q \leqslant p$ we finally obtain $q \leqslant p$ and $q \not\Vdash \phi$, i.e., the right-hand side is false.

LEMMA 11.6

$p \Vdash \forall v \ \phi(v)$ *is equivalent to* $\forall c \in C (p \Vdash \phi(c))$.

Proof. $[p] - F_{\forall v \ \phi(v)} = \bigcup_{c \in C}([p] - F_{\phi(c)})$ by 9.2. Using the σ-additivity of I we infer that this union belongs to I if and only if each of its members does.

LEMMA 11.7

$p \Vdash c \ \varepsilon \ \underline{m}$ *is equivalent to* $\forall q \leqslant p \ \exists r \leqslant q \ \exists n \in m$ $(r \Vdash c \approx \underline{n})$.

Proof. The left-hand side is equivalent to $\forall q \leqslant p$ $(q \Vdash c \ \varepsilon \ \underline{m})$ and hence to $\forall q \leqslant p$ $([q] - \bigcup_{n \in m} F_{c \approx \underline{n}} \in I)$ because $F_{c \varepsilon \underline{m}} = \bigcup_{n \in m} F_{c \approx \underline{n}}$. It follows that there exists an element n of m such that $[q] \cap F_{c \approx \underline{n}} \notin I$ and hence, by (C), there is a condition r such that $[r] - ([q] \cap F_{c \approx \underline{n}}) \in I$. This proves that $[r] \subseteq [q]$ and $[r] - F_{c \approx \underline{n}} \in I$, i.e., the left-hand side implies the right.

If the left-hand side is false then $[p] - F_{c \varepsilon \underline{m}} \notin I$ and hence, by (C), there is a q such that $[q] - ([p] - F_{c \varepsilon \underline{m}}) \in I$. It follows that $q \leqslant p$ and $[q] \cap F_{c \varepsilon \underline{m}} \in I$. Since $F_{c \approx \underline{n}} \subseteq F_{c \varepsilon \underline{m}}$ this proves that $[q] \cap F_{c \approx \underline{n}}$ is in I for each n in m and the same is true for each $r \leqslant q$.

LEMMA 11.8

$p \Vdash c \varepsilon \sigma$ is equivalent to $\forall q \leqslant p\ \exists r \leqslant q\ \exists s \geqslant r$ $(r \Vdash c \approx \underline{s})$.

Proof. We argue as in the previous proof using the equation $F_{c\varepsilon\sigma} = \bigcup_{s\in\underline{P}} ([s] \cap F_{c\approx\underline{s}})$.

LEMMA 11.9

If ϕ is a formula of L, $v \in Fr(\phi)$, $\alpha \in On_M$, $c' \in C_\alpha$ and $\gamma \in C_\alpha^{Fr(\phi)-\{v\}}$ then $p \Vdash c' \varepsilon c_{\alpha,\phi,\gamma}$ is equivalent to $p \Vdash \phi^{(\alpha)}(c',\gamma)$.

Proof. Putting $c = c_{\alpha,\phi,\gamma}$ we easily show that $F_{c'\varepsilon c} = F_{\phi^{(\alpha)}(c',\gamma)}$.

LEMMA 11.10

$p \Vdash V_\alpha c$ is equivalent to $\forall q \leqslant p\ \exists r \leqslant q\ \exists c' \in C_\alpha$ $(r \Vdash c \approx c')$.

Proof uses the same technique as 11.7 and the observation that $F_{V_\alpha c} = \bigcup_{c'\in C_\alpha} F_{c\approx c'}$.

LEMMA 11.11

If $c_1, c_2 \in C_{\alpha+1}$ then $p \Vdash c_1 \approx c_2$ is equivalent to

$$\forall c \in C_\alpha\{\forall q \leqslant p\ [(q \Vdash c \varepsilon c_1) \rightarrow \exists r \leqslant q\ (r \Vdash c \varepsilon c_2)]\ \&$$

$$\forall q \leqslant p\ [(q \Vdash c \varepsilon c_2) \rightarrow \exists r \leqslant q\ (r \Vdash c \varepsilon c_1)]\}.$$

Proof. It is immediate that $F_{c_1\approx c_2} = \bigcap_{c\in C} F_{\phi(c)}$ where ϕ is the formula $v \varepsilon c_1 \equiv v \varepsilon c_2$. From the σ-additivity of I it follows that $p \Vdash c_1 \approx c_2$ is equivalent to $\forall c \in C_\alpha (p \Vdash \phi(c))$. In order to bring the result to the desired form we express $\phi(c)$ by means of the connectives & and $\neg$ alone and obtain the sentence

$$\phi'(c) = \neg[\phi_1(c) \mathbin{\&} \neg\phi_2(c)] \mathbin{\&} \neg[\phi_2(c) \mathbin{\&} \neg\phi_1(c)]$$

where $\phi_i(c) = c \varepsilon c_i$ for $i = 1,2$. Since $\phi(c)$ and $\phi'(c)$ are logically equivalent (or more exactly: since $\phi(c)$ is just an abbreviation of $\phi'(c)$), the relations $p \Vdash \phi(c)$ and $p \Vdash \phi'(c)$ are equivalent. We now use Lemmas 11.4 and 11.5 and after easy transformations obtain the desired result.

LEMMA 11.12

If $c \in C_{\alpha+1}$ *and* $d \in C$ *then* $p \Vdash d \in c$ *is equivalent to* $\forall q_{\leqslant p} \exists r \leqslant q \, \exists c' \in C_\alpha[(r \Vdash d \approx c') \mathbin{\&} (r \Vdash c' \varepsilon c)]$.

Proof. Similar to that of 11.7 and uses the decomposition $F_{d\varepsilon c} = \cup[F_{d\approx c'} \cap F_{c'\varepsilon c}]$ where c' ranges over C_α.

12. Definability of the forcing relation.

We shall base our proof on the following theorem scheme on definability by transfinite induction. Let U be a subset of M and R a well founded relation which partially orders U. Let us assume that U and R are definable in M and that for each u in U the set of its R-predecessors $R(u) = \{v \in U : v \neq u \mathbin{\&} vRu\}$ belongs to M. Finally let H be a function definable in M which correlates an element of U with each pair a, A where $a \in U$, $A \in M$ and A is a function with domain $R(u)$. Under these assumptions there is a unique function G with domain U such that

$G(u) = H(u, G \upharpoonright R(u))$ for each u in U and this function is definable in M. (Note: $G \upharpoonright R(u)$ is the restriction of the function G to $R(u)$).

This theorem is but an inessential extension of the theorem on definitions by transfinite induction whose proof can be found in many textbooks of set theory. We shall not enter into details of this proof here.

We shall now prove the definability of the forcing relation. If ϕ is a formula of L which contains logical operators then either $\phi = \neg\psi$ or $\phi = \psi \,\&\, \theta$ or $\phi = \forall v\, \psi$ where v is a variable. If the relations $p \Vdash \psi(\gamma)$ and $p \Vdash \theta(\delta)$ are definable in M then so is the relation $p \Vdash \phi(\gamma)$ in view of Lemmas 11.4 - 11.6. Thus in order to verify assumption (D) it is sufficient to prove it for the case of atomic formulae. We shall establish a slightly stronger result:

LEMMA 12.1

The binary relation $p \Vdash \phi$ *where* $p \in P$ *and* ϕ *is a limited sentence of* RL *is definable in* M.

Proof. Let us consider pairs (p,ϕ) where $p \in P$ and ϕ is a limited sentence of RL. The set U of these pairs is definable in M. We order it partially by the following well founded relation R:

$$(p,\phi)R(q,\psi) \equiv \phi \prec \psi .$$

Let us put $G(p,\phi) = 0$ or 1 according as $p \Vdash \phi$ or $p \nVdash \phi$. In order to prove that the forcing relation is definable in M it is sufficient to show that the function G is definable in M

and we achieve this by showing that G satisfies a recursive equation $G(p,\phi) = H(p,\phi,G \restriction R(p,\phi))$ where H is a definable function. The proper choice of H becomes clear when we examine Lemmas 12.4 - 12.12. These lemmas show that the forcing relation $p \Vdash \phi$ can be reduced to some forcing relations between elements of P and limited sentences which precede ϕ with respect to the ordering $\prec$; thus these conditions can be expressed by means of the values of G limited to the set $R(p,\phi)$. E.g., if $\phi = \neg\psi$ then $G(p,\phi) = 0$ if and only if $\forall q \leq p\, [G(q,\psi) = 1]$. Accordingly we put $H(p, \neg\psi, A) = 0$ if and only if $\forall q \leq p\, (A(q,\psi) = 1)$. For other forms of ϕ the procedure is similar.

We can now give the exact definition of H. Let $a = (p,\phi)$; then $H(a,A)$ is defined for $a \in U$ and $A \in \{0,1\}^{R(a)} \cap M$.

If a has no R-predecessors then ϕ is one of the sentences $\underline{0} \varepsilon \underline{0}$, $V_0\underline{0}$, $\underline{0} \approx \underline{0}$ and we put $H(a,A) = 1$ in the first two cases and $= 0$ in the third.

If $\phi = \neg\psi$ then $H(a,A) = 0 \equiv \forall q \leq p\, (A(q,\psi) = 1)$;

If $\phi = \psi \,\&\, \theta$ then $H(a,A) = 0 \equiv A(p,\psi) = A(p,\theta) = 0$;

If $\phi = \forall_\alpha v\, \psi$ then $H(a,A) = 0 \equiv \forall c \in C_\alpha (A(p, \psi(c)) = 0)$;

If $\phi = c_1 \varepsilon c_2$ and $\rho(c_1) \geq \rho(c_2) = \alpha$ then $H(a,A) = 0 \equiv$

$$\forall q \leq p\, \exists r \leq q\, \exists c' \in C_\alpha [A(r, c_1 \approx c') = A(r, c' \varepsilon c_2) = 0];$$

If $\phi = c_1 \varepsilon c_2$ and $\rho(c_1) < \rho(c_2)$ and $c_2 = \underline{m}$ then

$$H(a,A) = 0 \equiv \forall q \leq p\, \exists r \leq q\, \exists n \in m\, (A(r, c_1 \approx \underline{n}) = 0);$$

If $\phi = c_1 \varepsilon c_2$ and $\rho(c_1) < \rho(c_2)$ and $c_2 = \sigma$ then

$$H(a,A) = 0 \equiv \forall q \leq p\, \exists r \leq q\, \exists s \geq r\, (A(r, c_1 \approx \underline{s}) = 0);$$

If $\phi = c_1 \varepsilon c_2$ and $\rho(c_1) < \rho(c_2)$ and $c_2 = c_{\alpha,\phi,\gamma}$ then

$$H(a,A) = 0 \equiv A(p,\psi^{(\alpha)}(c_1\gamma)) = 0;$$

If $\phi = c_1 \approx c_2$ and $\max(\rho(c_1), \rho(c_2)) = \alpha + 1$ then

$$H(a,A) = 0 \equiv \forall c \in C_\alpha \{\forall q \leqslant p\ [A(q,c\ \varepsilon\ c_1) = 1 \vee \exists r \leqslant q\ A(r,c\ \varepsilon\ c_2) = 0]\ \&\ \forall q \leqslant p\ [A(q,c\ \varepsilon\ c_2) = 1 \vee \exists r \leqslant q\ A(r,c\ \varepsilon\ c_1) = 0]\};$$

If $\phi = V_\alpha c$ then $H(a,A) = 0 \equiv \forall q \leqslant p\ \exists r \leqslant q\ \exists c' \in C_\alpha$

$$(A(r,c \approx c') = 0).$$

The function H is of course definable in M. Using Lemmas 12.4 - 12.12 we prove that $G(a) = H(a, G \restriction R(a))$ for each a in U. Hence G is definable in M and so is the forcing relation because $p \Vdash \phi \equiv G(p,\phi) = 0$ whenever ϕ is a limited sentence. Thus condition (D) is verified.

13. *Additional remarks*

Let $D \subseteq P$ be a set dense in P i.e., such that for every p in P there is a q in D such that $q \leqslant p$. We shall say that D is dense in P under p if $\forall q \leqslant p\ \exists r \in D\ \ r \leqslant q$.

In theorem 9.9 we established the existence of a sequence $\{p_n\}$ of conditions which has the property that the set $F = \{\phi : \exists n\ p_n \Vdash \phi\}$ is closed. We want to characterize sequences with this property.

THEOREM 13.1

If $\{p_n\}_{n\in\omega}$ *is a sequence such that the set* $F = \{\phi : \exists n\ p_n \Vdash \phi\}$ *is closed then the filter* X *generated by* $\{p_n\}$ *has common elements with every set* D *which belongs to* M *and is dense in* P.

Proof. Let $D \in M$ be a dense set. If the sentence $\exists v\,[(v \varepsilon \underline{D}) \,\&\, (v \varepsilon \sigma)]$ belongs to F then there is an element p of M such that $\underline{p} \varepsilon \underline{D} \in F$ and $\underline{p} \varepsilon \sigma \in F$. It follows that for some integer n, $p_n \Vdash \underline{p} \varepsilon \underline{D}$ and $p_n \Vdash \underline{p} \varepsilon \sigma$. The first relation implies $p \in D$ and the second $p_n \leqslant p$ (see 11.8). Hence $p \in D \cap X$.

We shall now show that the assumption $\exists v\,[(v \varepsilon \underline{D}) \,\&\, (v \varepsilon \sigma)] \notin F$ leads to a contradiction. This assumption implies that the sentence $\forall v\,\neg[(v \varepsilon \underline{D}) \,\&\, (v \varepsilon \sigma)]$ belongs to F and thus is forced by a condition p_n from the initially given sequence. Thus for each constant c the condition p_n forces the sentence $\neg[(c \varepsilon \underline{D}) \,\&\, (c \varepsilon \sigma)]$. We choose for c the constant $\underline{q}$ where q is an element of D such that $q \leqslant p_n$. Using Lemma 11.5 we obtain $q \nVdash (\underline{q} \varepsilon \underline{D}) \,\&\, (\underline{q} \varepsilon \sigma)$, i.e., either $q \nVdash \underline{q} \varepsilon \underline{D}$ or $q \nVdash \underline{q} \varepsilon \sigma$. Both these alternatives are clearly false.

A filter X is called generic if $X \cap D \neq \emptyset$ for each set D which is dense in P and belongs to M (more exactly X is called generic in P over M). Theorem 13.1 can thus be expressed as follows: If $\{p_n\}_{n\in\omega}$ is a sequence such that the set $\{\phi : \exists n\ p_n \Vdash \phi\}$ is closed then the filter generated by the p_n's is generic. We shall prove that also the converse of this theorem is true. First we need a lemma:

LEMMA 13.2

A generic filter intersects every set $D \subseteq P$ *which belongs to* M *and is dense under* p *where* p *is any element of the filter.*

Proof. Put $D' = D \cup \{q \in P : q$ is incompatible with $p\}$. In view of the separability of P the set D' is dense in P and hence if X is generic then $X \cap D' \neq \emptyset$. Since no element of X is incompatible with p, we obtain $X \cap D \neq \emptyset$.

THEOREM 13.3

If X *is generic then the set* $F = \{\phi : \exists p \in X\ p \Vdash \phi\}$ *is closed.*

Proof. (1) F is consistent. Otherwise there would be a finite set $\phi_1,\ldots,\phi_n$ of sentences in F such that the conjunction ϕ of these sentences is inconsistent, i.e., has no model. By assumption each ϕ_j is forced by a condition p_j in X. Since X is a filter we obtain a condition p in X such that $p \leq p_j$ for each $j \leq n$ and so $p \Vdash \phi$. Thus $[p] - F_\phi \in I$, therefore $[p] \cap F_\phi \neq \emptyset$ and we obtain a contradiction because for each Y in F_ϕ the family M[Y] is a model of ϕ.

(2) F is complete. Let ϕ be a sentence and $D = \{p \in P : p \Vdash \phi$ or $p \Vdash \neg\phi\}$. In view of the definability of the forcing relation we have $D \in M$. We shall show that D is dense in P. For let q be any condition. By 11.5 if $q \nVdash \neg\phi$ then there is a condition $p \leq q$ such that $p \Vdash \phi$. Hence either $q \in D$ or some extension p of q is in D.

Since X is generic we obtain now $X \cap D \neq \emptyset$; if p belongs to this intersection, then either p forces ϕ or p forces

$\neg\phi$ whence either ϕ or $\neg\phi$ belongs to F.

(3) The axioms 1 - 12 given in Section 8 belong to F. This is so because these axioms are true in all models M[Y] where Y is any filter; hence they are forced by *any* conditions.

(4) F satisfies the postulates (ii) - (vi). Since the verification is practically the same for all the postulates we shall give the proof only for the postulate (ii). Thus let us assume that $\exists v_n\ \phi \in F$, i.e., $p \Vdash \exists v_n\ \phi$ for some p in X. Since $\exists v_n$ is an abbreviation of $\neg\forall v_n\neg$ we can apply Lemmas 11.5 and 11.6 and obtain $\forall q \leq p\ \exists r \leq q\ \exists c \in C\ r \Vdash \phi(c)$. This means that the set $D = \{r \in P : \exists c \in C\ r \Vdash \phi(c)\}$ is dense under p and so, since this set belongs to M, we obtain that there is a condition r in $D \cap X$. Hence there is a constant c such that $r \Vdash \phi(c)$ which proves that $\phi(c) \in F$. The postulate (ii) is thus verified.

Theorems 13.1 and 13.3 suggest an alternative method of constructing models. We start with the definition of forcing and establish first of all Lemmas 11.1 - 11.12. Then we define generic filters and prove their existence essentially as in the proof of the Baire theorem. Next we establish Theorem 13.2 obtaining a closed set. Finally we prove that each closed set determines a reducible filter X as we did in Section 8.

This alternative method is essentially the one which was used by Cohen. Most authors follow Cohen by defining forcing from the start by transfinite induction (in the model M). This allows then to avoid the cumbersome verification of condition (D). The only defect of this method is that it is not easy for the beginner to grasp the intuitive meaning of the forcing relation.

The connection between forcing and the concept of meager sets was discovered by Takeuti and Ryll-Nardzewski and we exploited their ideas in the proofs given above.

14. *Preservation of cardinals*

If $\alpha \in M$ is an ordinal then we say that α is a cardinal of M if there is no element f of M which is a mapping of a smaller ordinal onto α. One can show by examples that a cardinal of M need not be a cardinal of $M[X]$.

We shall derive a sufficient condition for a cardinal of M to remain a cardinal of $M[X]$.

DEFINITION

We denote by $\theta_M(P)$ the least ordinal α of M such that for each set $Q \subseteq P$ consisting of mutually incompatible conditions and such that $Q \in M$ there is in M a one-one mapping of Q into α.

LEMMA 14.1

If $M \Vdash ZFC$ *then* $\theta_M(P)$ *exists.*

Proof. We can formalize in ZFC the proof that for each partially ordered set there is a least cardinal larger than or equal to the cardinal of any set of mutually incompatible elements of P.

LEMMA 14.2

If X *is a generic filter and* α *is a cardinal of* M, $M \Vdash ZFC$ *and* $\alpha > \theta_M(P)$ *then* α *is a cardinal of* $M[X]$.

Proof. Let us assume that there is in $M[X]$ a function f and an ordinal $\beta < \alpha$ such that f maps β onto α. Expressing these facts in RL we obtain a formula $\phi(f,\underline{\alpha},\underline{\beta})$: Funct(f) & (Dom(f) = $\underline{\beta}$) & (Rg(f) = $\underline{\alpha}$) which is true in $M[X]$. Denoting by c a constant such that $c^*[X] = f$ we infer from the assumption that X is generic that there is a condition p_0 in X satisfying $p_0 \Vdash \phi(c,\underline{\alpha},\underline{\beta})$.

Consider now the set

$$S = \cup\{Z_\xi : \xi < \beta\} \text{ where } Z_\xi = \{\eta : \exists p \leqslant p_0$$

$$p \Vdash \exists v\ [\Pi(v,\underline{\xi},\underline{\eta})\ \&\ v \in c]\} \cap \alpha.$$

From the definability of $\Vdash$ we see that $Z_\xi \in M$.

The cardinal number of Z_ξ (calculated in M) is $\leqslant \theta_M(P)$. To see this we correlate (using the axiom of choice) a condition $p \leqslant p_0$ to each η in Z_ξ so that $p \Vdash \underline{\langle \xi,\eta \rangle} \in c$. Since $p_0 \Vdash$ Funct(c) it cannot be the case that two compatible p_1, p_2 be correlated to two different ordinals η_1,η_n. Hence the set of conditions correlated with elements of Z_ξ consists of mutually incompatible conditions and hence its cardinal number (in M) is $\leqslant \theta_M(P)$. Since $\beta < \alpha$ and $\theta_M(P) < \alpha$ it follows that the cardinal number of S is $< \alpha$. On the other hand $\alpha \subseteq S$ because Rg(f) = α and thus for each η in α there is a ξ in β and a p in X such that $p \Vdash \underline{\langle \xi,\eta \rangle} \in c$. Lemma 14.2 is thus proved.

15. *The independence of* CH

Let $M \Vdash$ ZFC, $\alpha \geqslant \omega$, $x \in M$. We take as P the set of finite functions p such that Dom(p) $\subseteq \alpha \times \omega$, Rg(p) $\subseteq \{0,1\}$.

P is obviously an element of M. We order P by convention: $p \leq q$ if and only if $p \supseteq q$. All the assumptions which we made in Section 1 are satisfied by P. In particular p, q are compatible if and only if they coincide on the intersection of their domains.

LEMMA 15.1

$$\theta_M(P) = \omega.$$

The proof is due to Cohen but the theorem was already proved in 1941 by Marczewski. In order to prove the lemma we formalize the following reasoning in ZFC.

We consider finite functions with values in $\{0,1\}$ and with domains $\subseteq A$ where A is infinite (in our case $A = \alpha \times \omega$). Let us assume that there is a non-denumerable set R_1 of mutually incompatible such functions. R_1 can be decomposed into a denumerable union of sets, two functions being included in the same set if their domains have the same number of elements. One of these sets is non-denumerable; to save notation we assume that all the functions in R_1 have domains of power exactly k. Let $p_1 \in R_1$. If $q \in R_1$ and $p_1 \neq q$, then p_1 and q are incompatible and this can happen only if there is an element $t = t_q$ in the domain of p_1 such that $t \in \mathrm{dom}(q)$ and $p_1(t) \neq q(t)$. Now there are only finitely many elements t in $\mathrm{dom}(p_1)$ and non-denumerably many q's. Hence there is a non-denumerable family $R_2 \subseteq R_1$ and an element $t_1 \in \mathrm{dom}(p_1)$ such that for each $q \in R_1$ the element t_1 is in $\mathrm{dom}(q)$ and $p_1(t_1) \neq q(t_1)$. Let $p_2 \in R_2$. Hence $\mathrm{dom}(p_2)$ contains t_1. Again we see that there are non-denumerably many elements $q \in R_2$ such that for some fixed t_2, $t_2 \in \mathrm{dom}(p_2)$, $q(t_2) \neq p_2(t_2)$. It cannot be the case that $t_2 = t_1$ because we would then have $q(t_2) = 1 - p_2(t_2) = 1 - p_2(t_1) = 1 - (1 - p_1(t_1))$ since $p_2(t_1) = 1 - p_1(t_1)$. But this is impossible because $q(t_2)$, i.e. $q(t_1)$, is different from $p_1(t_1)$. Thus we see that $\mathrm{dom}(p_2)$ has at least 2 elements t_1 and t_2. Continuing this reasoning we obtain a p_3 and a non-denumerable subset R_3 of R_2 such that p_3 has at least 3 elements in its domain. After $k + 1$ steps we arrive at a p_{k+1} in $R_k \subseteq R_1$ with $k+1$ elements in its domain which contradicts our assumption.

THEOREM 15.2

There are models in which CH *is false.*

Proof. Let $M \Vdash$ ZFC be denumerable; take any cardinal α of M which is greater than the first uncountable cardinal of M. Define P as in the lemma above and let X be generic in P. We are going to prove that $M[X] \Vdash \neg$CH. Since $M[X]$ and M have the same cardinals it is sufficient to show that $M[X]$ has an element f which is a function with domain α, with range $\subseteq 2^\omega \cap M[X]$ and which is an injection.

We obtain f by taking the union $\Phi = \cup X$ which is a mapping of $\alpha \times \omega$ into $\{0,1\}$ and putting $f(\xi) = \Phi_\xi$ where $\Phi_\xi(n) = \Phi(\xi,n)$. Obviously $\Phi \in M[X]$ and so $f \in M[X]$. It remains to prove that f is an injection, i.e. $f(\xi) \neq f(\eta)$ for $\xi \neq \eta$. Let us assume that this is not the case, i.e. that there are $\xi, \eta < \alpha$ such that $\xi \neq \eta$ and $\Phi(\xi,n) = \Phi(\eta,n)$ for each n. The following formula of RL expresses this fact (i.e. is true in $M[X]$):

$$\forall v_0\ \forall v_1\ \{v_0 \,\varepsilon\, \sigma \rightarrow [\langle \underline{\xi}, v_1, \underline{0} \rangle \,\varepsilon\, v_0 \equiv \langle \underline{\eta}, v_1, \underline{0} \rangle \,\varepsilon\, v_0]\}$$

(as before we assumed here that ordered pairs and triplets can be defined by formulae of RL). It will be more convenient to write this formula as

(*) $\forall v_0 \forall v_1 \neg\{(v_0 \in \sigma) \& \neg\psi\}$

where ψ is the formula $\langle \underline{\xi}, v_1, \underline{0}\rangle \in v_0 \equiv \langle \underline{\eta}, v_1, \underline{0}\rangle \in v_0$. Since (*) is assumed to be true in $M[X]$ there is a p in X which forces this formula. Using Lemmas 11.4 and 11.5 we infer that for each q in P and each n in ω

(**) $p \Vdash \neg\{(\underline{q} \in \sigma) \& \neg\psi(\underline{q}, \underline{n})\}$.

We now notice that the domain of p is finite and thus there is n such that neither $\langle \xi, n\rangle$ nor $\langle \eta, n\rangle$ are in dom(p). Adding $\langle \xi, n, 1\rangle$ and $\langle \eta, n, 0\rangle$ to p we obtain a condition $q \leq p$. Apply now to (**) the Lemma 11.4. We obtain

$$q \nVdash [\underline{q} \in \sigma \& \neg\psi(\underline{q}, \underline{n})]$$

i.e. $q \nVdash (\underline{q} \in \sigma)$ or $q \nVdash \neg\psi(\underline{q}, \underline{n})$ which is equivalent to $q \nVdash (\underline{q} \in \sigma)$ or $\exists r \leq q \ r \Vdash \psi(\underline{q}, \underline{n})$. The first part of this disjunction is obviously false. The second is false too because for each $r \leq q$ and each Y in $[r]$ the truth value of $\langle \underline{\xi}, \underline{n}, \underline{0}\rangle \in \underline{q}$ in $M[Y]$ is "false" and that of $\langle \underline{\eta}, \underline{n}, \underline{0}\rangle \in \underline{q}$ is "true" and so $r \nVdash \psi(\underline{q}, \underline{n})$.

BIBLIOGRAPHY

COHEN, P. J.

[66] Set theory and the continuum hypothesis, Benjamin, 1966.

GÖDEL, K.

[40] The consistency of the continuum hypothesis, Princeton University Press, 1940.

KURATOWSKI, K.

[66] Topology, Vol. 1. Academic Press and PWN, 1966.

* * *

Department of Mathematics, University of Warsaw, Warsaw, Poland.

LOGIC AND FOUNDATIONS

A. Nerode[1]

0. Let us begin by reassuring those in the audience unconnected with logic (or even mathematics) that our topic is historical rather than technical. By logic we intend the older narrow meaning, the study of the rules of reasoning. By foundations we intend the nineteenth century meaning, the study of the construction and axiomatic defining properties of mathematical objects and ideas (what is an irrational? a transfinite ordinal? a set? an infinitesimal?).

Revolutionary ideas are likely to be difficult to express in terminology previously available. They are likely to be confused with other ideas in a contradictory fashion. They are likely to be met with unsympathetic disbelief by minds of lesser stature than those of the discoverers. This is as true of mathematics as of any of the other sciences – perhaps even more so due to the standard of demonstrative precision associated with mathematics.

The term foundations has lately been applied primarily to the twentieth century problems arising from trying to understand the concepts of class and set after the Burali-Forti paradox. This emphasis reflects the fact that whatever is of current research interest tends to be regarded as the whole field. But this is only one in a

---

[1]Opening address at the Conference.

long historic sequence of mathematical foundations questions, and by no means the most important for mathematical practice.

Our purpose is to emphasize that "logic and foundations" investigations have accompanied many of the revolutionary ideas of historic mathematics and have profoundly affected mathematical practice by contributing constructions or defining properties for the newly discovered ideas in a form easily assimilated and used by the working mathematician. These constructions or defining properties are often dismissed as obvious by later generations with no appreciation that many generations of foundational thought may have separated the first bloom of an idea from its' characterization in a form assimilable in deductive mathematics. When foundations are absorbed into mathematics proper, they vanish.

We admit a personal prejudice against those who confuse the knife and fork with the food – namely those who put axioms which attempt to capture the essential features of ideas on the same plane as the routine rules of reasoning used to manipulate them.

We choose examples familiar to everyone to reiterate the points made above.

1. The Pythagorean School (ca. 550 B.C.) revealed the irrational numbers. Rational numbers, quotients of whole numbers, arise geometrically by subdivision of unit intervals into an integral number of equal parts. The operations on rationals were easily developed by the Greeks. If a right triangle is constructed with shorter legs of length one, the "Pythagorean theorem" tells us that the length ℓ of the hypotenuse has $\ell^2 = 2$; and an elegant argument attributed to

Pythagoras shows that the assumption ℓ is rational leads to a contradiction. So if geometrically constructed segments are to have length, more numbers than the rationals are needed. Thus were the irrational numbers revealed. But then what is an irrational number? What are their properties? This question casts a shadow on the coherence of Greek geometry, illuminated by Eudoxus (ca. 370 B.C.). He (by tradition) put the keystone in the arch of Greek geometry, as developed in Plato's academy and finally compiled in Euclid's Elements (ca. 330 B.C.). In modern terms what he did was to formulate a necessary and sufficient condition for two Dedekind sections to determine the same real number. What he and Archimedes used was a necessary and sufficient condition for two sequences of rational numbers to have the same limit (reformulated in the nineteenth century by Cauchy as the condition for two Cauchy sequences of rationals to be equivalent). Some people express Eudoxus' contribution in algebraic terms by saying that before him the Greeks had an ordered field, afterwards an archimedean ordered field. But this misses the point that he produced a precise and usable theory of limits, exploited later by Archimedes. In Euclid's Elements and Archimedes the Eudoxian method is used with virtually the precision of the Weierstrass ε-δ method, but in cumbersome geometric terms. In modern terms, they introduce limits of sequences after showing them to be Cauchy. The sequences are obtained by geometric constructions designed to rectify curves, perform quadratures of regions, etc.; the limits are the desired lengths and areas. But, unlike in the nineteenth century work, no single equivalent of a completeness or continuity axiom appears. The apparatus for dealing with arbitrary Cauchy sequences was present, the notion of an arbitrary sequence absent. Of course for the Dedekind construction and characterization of the reals, the general notion is needed. On the other hand a late twentieth century constructivist might commend Eudoxus and Euclid for introducing an irrational only when the need arose. In any case two fundamental ideas, limit and archidemean ordering, had been made part of axiomatic mathematics.

What was the contribution of Euclid to logic? Euclid tried to derive geometry from fixed axioms. In this he was not wholly successful, as the axioms were not sufficient. But there would have been no point whatever if the means allowed in demonstrations introduced unintended additional premises. It was therefore necessary to check at every stage that only valid rules of logic were used. The result is that the reasoning in Euclid is highly stylized, the rules used few. The rules of reasoning are merely used, not formulated there as general principles. Nevertheless, for 2000 years the best way to learn logic was by studying Euclid. Why? Geometry deals with the relation of incidence (point on line, etc.). The rules of logic used in Euclid are the rules of the logic of relations (predicate logic and more). But the *systems* of logic of the ancient Greek philosophers appear to have dealt only with properties of things, not relations between things, and did not encompass the reasoning rules actually used by mathematicians of the time, much less those of philosophers.

Mathematicians today use no rules of logic not familiar from Euclid. The implicitly agreed upon rules of logic used by mathematicians outstripped any formal system of logic until Frege (1879). His system, the culmination of the work of De Morgan, Boole and Pierce, simply incorporates in generality the rules of Euclid. Perhaps this is why mathematicians regard formal logic as obvious; they have been speaking it all their lives. Nevertheless, making logic a branch of mathematics was an important achievement.

2. Infinitesimals were used by early workers in the calculus. In the hands of Leibniz (1646-1716) they resulted in the simple algebraic rules of calculation for derivatives and integrals that we use today. His successors were, for nearly three hundred years, unable to formulate fixed principles from which demonstrations using infinitesimals could be deduced. They gradually fell into disrepute among rigorous mathematicians, but never lost their power to suggest results to both pure and applied mathematicians. In the middle of the nineteenth century Weierstrass showed how to avoid them entirely for calculus, basing it solely on the real numbers. In 1960 Abraham Robinson (1919-1974) fully justified the infinitesimal method. He established reals and infinitesimals as part of a larger field R^*. The specific field used was discovered by other logicians earlier, but to Robinson belongs the sole credit for the realization that it was a suitable context for Leibniz' theory. A fundamental reason why this field R^* succeeded where previous ones had failed was the existence of "transfer principles". For example, each function of reals naturally extends to a function on the larger field R^*, and each predicate logic statement true of the reals transfers to a predicate logic statement true of R^*. A reading of Leibniz shows that he repeatedly *uses* transfer principles correctly, but where he tries to explain in general what will transfer, he fails miserably. Predicate logic statements transfer, most others do not, and the distinction was unknown at the time. Imitators of Leibniz without his intuition would get nonsense; and many did.

Robinson's book *Non-standard Analysis* (1966) gives the only balanced account of the history of calculus, being the only one written by someone who knew that infinitesimals exist. All other calculus histories act as if it were a great achievement to eliminate infinitesimals. This is nonsense. The Weierstrass achievement was to obtain calculus by demonstration from fixed principles. The naturalness of the infinitesimals was sacrificed for rigour.

The early history of infinitesimals is not mentioned in Robinson, but it is instructive. In 1906 a tenth century copy of a lost paper of Archimedes was discovered in Constantinople. This is in Heath's book on Archimedes and is called *The Method.* Here Archimedes gave away his "trade secrets", the method by which he discovered his rectifications and quadratures. In his other papers they are demonstrated without any hint as to how the formulas were discovered. What we find here, remarkable to behold, is the infinitesimal method. He makes no claim for rigour, merely recommending it as a research tool. But with Robinson's infinitesimals many arguments can be directly justified without recourse to Eudoxus (or, equivalently, Weierstrass, whichever you prefer). So Archimedes too used infinitesimals well, but did not see how to arrange demonstrations using them. (Some historians guess that all traces of infinitesimals have been deliberately erased in Greek geometry due to the foundations problem.)

Last, we remark that many of the arguments of Cavalieri, Kepler, Wallis, etc. for rectification and quadrature used infinitesimals in a way perfectly justifiable via Robinson. Thus it appears that infinitesimals have been used throughout mathematical history as a tool for discovery and research, despite their disreputable character. Now that they have been made respectable we would hope that a serious historical study of their rôle will be made.

3. Cantor (1845-1918) saw transfinite cardinals and ordinals as clearly as his arch-critic Kronecker saw the integers. His exposition was as informal as that of Diophantus for ancient number theory. It was very hard to follow. He spoke of two principles of generation. One generates for every set of ordinal numbers the next larger; the other generates the set of all subsets of a set A from the set A. In 1897 Burali-Forti published a famous paper in which he attempted to prove a certain proposition of Cantor's theory by assuming it false and getting a contradiction. But it was observed that the assumption that the proposition was false was never used. So a purported contradiction in Cantor's theory appeared. As a result, many regarded Cantor's set theory as doubtful, and the 20th century is replete with solutions to this "paradox". But in fact the contradiction arose from the assumption that there is a set of *all* ordinals, which explicitly contradicts Cantor's generation principle.

This certainly showed a need for fixed principles from which to deduce Cantor's easily misunderstood set theory. He himself used a few simple constructions of new sets from old, and these were more or less adequately listed by Zermelo (1908). (There was an omission remedied by Skolem and Fraenkel, 1920.) These gave a fixed set of principles which anyone can understand and use for the Cantorian theory as it then existed.

It seems to me that the furore over the paradoxes at the turn of the century was really over a different issue. Two properties are said to have the same extension (or determine the same class) if they apply to the same "objects". Frege (1893) introduced the extensions of all mathematical properties and treated them as if they were sets: that is, subject to Cantor's rules of set construction. The paradoxes show that they are not.

The theory of extensions of properties is now understood to be a subject different from Cantor's set theory, and is still quite undeveloped.

4. In the old days all constructions which gave complicated systems from simpler ones were regarded as foundations. (For example, reals from rationals, rationals from integers, polynomial extensions from fields, algebraic extensions from polynomial extensions). Now these are wholly absorbed in the working apparatus of all mathematicians. A paradigm of our examples would be Pythagoras, (Leibniz, Cantor) intuit and exploit fundamental ideas of irrationals (infinitesimals, transfinite numbers). Much confusion results till Eudoxus,(Robinson, Zermelo) make clear the principles that govern the new notions. At this point there is a sigh of relief, and mathematical life goes on.

We have omitted out of clarity the obvious references to famous mathematicians and philosophers who dismissed irrationals, infinitesimals and transfinite numbers as meaningless. We cannot resist an anecdote told us by a very well-known mathematician. In the early 1930's he had the technical apparatus and a good idea how to do Leibniz' calculus. His illustrious advisers told him that infinitesimals had been shown to be useless. He dropped the subject So much for the opinions of greybeards!

In this talk we have not advertised the ideas of twentieth century logic, such as recursive functions, saturation, constructibility or forcing. But, being current, they already receive enough emphasis.

* * *

Department of Mathematics, Cornell University, Ithaca, N.Y.

CHURCH-ROSSER THEOREMS FOR REPLACEMENT SYSTEMS

John Staples

Replacement systems with the Church-Rosser property promise to be of great interest for the theory of parallel programming languages; particular systems have long been of interest in theories of combinatory logic and lambda conversion. This paper reviews known methods for proving the Church-Rosser property for general replacement systems and adds some new results. Finally some open problems are listed.

1. *Introduction*

Although Church-Rosser theorems for particular systems have been known for several decades the problem in the generality considered here has been studied only more recently. We define after Rosen [73] a replacement system $\mathcal{B} = (B,\Rightarrow)$ on a set B to be the set B together with a binary relation $\Rightarrow$ on B. We write $\Rightarrow^*$ for the reflexive, transitive closure of $\Rightarrow$.

The Church-Rosser property for such a replacement system is the following:

(CR) *for all* $a,b,c \in B$ *such that* $a \Rightarrow^* b$, $a \Rightarrow^* c$, *there is* $d \in B$ *such that* $b \Rightarrow^* d$, $c \Rightarrow^* d$.

A replacement system is said to be Church-Rosser, or CR, if it has this property. The CR problem for a system $\mathcal{B}$ is whether

$\mathcal{B}$ is CR.

We are interested in giving conditions on $\mathcal{B}$ which ensure that it is CR. Relatively little has been done on this problem for replacement systems in general; Rosen [73] gives, among others, the following three results.

1.1 *Suppose that there is a binary relation* $\Rightarrow_1$ *on* B *such that* $\Rightarrow_1^* = \Rightarrow^*$ *and such that for all* $a,b,c \in B$, *if* $a \Rightarrow_1 b$, $a \Rightarrow_1 c$ *then there is* $d \in B$ *such that* $b \Rightarrow_1 d$, $c \Rightarrow_1 d$; *then* $\mathcal{B}$ *is* CR.

1.2 Say that two replacement systems $\mathcal{B}_1 = (B, \Rightarrow_1)$ and $\mathcal{B}_2 = (B, \Rightarrow_2)$ over the same set B commute with each other if for all $a,b,c \in B$ such that $a \Rightarrow_1^* b$, $a \Rightarrow_2^* c$, there is $d \in B$ such that $b \Rightarrow_2^* d$, $c \Rightarrow_1^* d$. Then

If $\{\mathcal{B}_\alpha : \alpha \in A\}$ *is a family of replacement systems over the same set* B, *say* $\mathcal{B}_\alpha = (B, \Rightarrow_\alpha)$, *and if for all* $\alpha, \beta \in A$ $\mathcal{B}_\alpha$ *commutes with* $\mathcal{B}_\beta$, *then their union* $\mathcal{B} = (B, \bigcup_{\alpha \in A} \Rightarrow_\alpha)$ *is* CR.

1.3 Note that a replacement system is CR just when it commutes with itself; hence the following result also gives a Church-Rosser theorem. Write $\Rightarrow^=$ for the reflexive closure of a binary relation $\Rightarrow$.

A sufficient condition for systems $(B, \Rightarrow_1)$, $(B, \Rightarrow_2)$ *to commute is that for all* $a,b,c \in B$ *such that* $a \Rightarrow_1 b$, $a \Rightarrow_2 c$ *there is* $d \in B$ *such that* $b \Rightarrow_2^* d$ *and* $c \Rightarrow_1^= d$.

As Rosen observes, 1.2 and 1.3 had previously been given by Hindley [64].

1.4 Another well-known and useful sufficient condition for systems $(B, \Rightarrow_1)$ and $(B, \Rightarrow_2)$ to commute is:

for all $a,b,c \in B$ *such that* $a \Rightarrow_1 b$, $a \Rightarrow_2^* c$ *there is* $d \in B$ *such that* $b \Rightarrow_2^* d$, $c \Rightarrow_1^* d$.

1.5 The result just given may be refined as follows.

If there is a relation $\Rightarrow_+$ *on* B *such that* $\Rightarrow_1 \subseteq \Rightarrow_+ \subseteq \Rightarrow_1^*$ *and such that for all* $a \Rightarrow_+ b, a \Rightarrow_2 c$ *there is* d *such that* $c \Rightarrow_+ d$ *and* $b \Rightarrow_2^* d$, *then* $(B,\Rightarrow_1)$ *and* $(B,\Rightarrow_2)$ *commute.*

This result is the basis of the Mitschke-Rosen method of section 4, though the value of that method lies in the techniques which are used to define the relation $\Rightarrow_+$.

1.6 Rosen [73] gives the following condition, weaker than commutativity, for the union of two CR systems to be CR. Say that $(B,\Rightarrow_1)$ requests $(B,\Rightarrow_2)$ if for all $a,b,c \in B$ such that $a \Rightarrow_1^* b$, $a \Rightarrow_2^* c$ there are $d,e \in B$ such that $b \Rightarrow_2^* d$, $c \Rightarrow_1^* e$, $e \Rightarrow_2^* d$. Clearly a system is CR just when it requests itself. Now:

if $(B,\Rightarrow_1)$, $(B,\Rightarrow_2)$ *are* CR *and the former requests the latter then their union is* CR.

Rosen also observes that

1.7 *A sufficient condition for* $(B,\Rightarrow_1)$ *to request* $(B,\Rightarrow_2)$, *where the latter is* CR, *is that for all* $a,b,c \in B$ *such that* $a \Rightarrow_1^* b$, $a \Rightarrow_2 c$ *there are* $d,e \in B$ *such that* $b \Rightarrow_2^* d$, $c \Rightarrow_1^* e$, $e \Rightarrow_2^* d$.

1.8 The hypotheses of Rosen's result 1.6 can be weakened. Instead of requiring that $(B,\Rightarrow_1)$ be CR we proceed as follows. Define $\Rightarrow_c$ to be the composite of $\Rightarrow_1^*$ and $\Rightarrow_2^*$; that is $a \Rightarrow_c b$ holds if and only if there is d such that $a \Rightarrow_1^* d$, $d \Rightarrow_2^* b$. Our hypothesis on $(B,\Rightarrow_1)$ is that for all $a \Rightarrow_1^* b$, $a \Rightarrow_1^* e$ there is d such that $b \Rightarrow_c d$, $e \Rightarrow_c d$.

To prove from this weaker hypothesis that the union is CR we observe that $\Rightarrow_c^*$ is $(\Rightarrow_1 \cup \Rightarrow_2)^*$, then we prove that if $a \Rightarrow_c b$, $a \Rightarrow_c e$ then there is d such that $b \Rightarrow_c d$, $e \Rightarrow_c d$. The result then follows by 1.1.

By hypothesis there are b', e' such that $a \Rightarrow_1^* b'$, $b' \Rightarrow_2^* b$, $a \Rightarrow_1^* e'$, $e' \Rightarrow_2^* c$. Hence there are d', f', g' such that $b' \Rightarrow_1^* d'$, $d' \Rightarrow_2^* g'$, $c' \Rightarrow_1^* f'$, $f' \Rightarrow_2^* g'$. Since $(B,\Rightarrow_1)$ requests $(B,\Rightarrow_2)$ there are p, q such that $b \Rightarrow_1^* p$, $p \Rightarrow_2^* q$, $d \Rightarrow_2^* q$. Similarly there are p', q' such that $e \Rightarrow_1^* p'$, $p' \Rightarrow_2^* q'$, $f' \Rightarrow_2^* q'$. Now as $(B,\Rightarrow_2)$ is CR there are s, s' such that $q \Rightarrow_2^* s$, $g' \Rightarrow_2^* s$, $q' \Rightarrow_2^* s'$, $g' \Rightarrow_2^* s'$, hence there is also d such that $s \Rightarrow_2^* d$, $s' \Rightarrow_2^* d$. As $b \Rightarrow_1^* p$, $p \Rightarrow_2^* q$, $q \Rightarrow_2^* s$, $s \Rightarrow_2^* d$ we have $b \Rightarrow_c d$, and $e \Rightarrow_c d$ similarly as required.

1.9 The above results may be applied to simplify CR problems but it usually remains nontrivial to solve the simplified problems. A similar situation occurs with the abstract CR theorems of Hindley [64,69] and Schroer [65]; it is nontrivial to show that their complex hypotheses are satisfied. See Hindley [74]. The same criticism can be made of the generalisation given in section 4 of the Mitschke-Rosen method. It is these problems and the hope of sharper results which keep alive interest in Church-Rosser theorems in less general contexts such as subtree replacement systems.

2. *Acyclic systems*

The following condition is clearly necessary for a replacement system $(B,\Rightarrow)$ to be CR:

(*) *for all* $a,b,c \in B$ *such that* $a \Rightarrow b$, $a \Rightarrow c$ *there is* d *such that* $b \Rightarrow^* d$, $c \Rightarrow^* d$.

Furthermore the condition is often easy to verify. It is not however a sufficient condition for the system to be CR, as Hindley [74] has shown by the following simple counterexample. Take B to be a four-element set $\{a,b,c,d\}$ and define $x \Rightarrow y$ if and only if $(x,y) = (b,a)$, (b,c), (c,b) or (c,d). The system is indicated by the directed graph in figure 1.

In this example the cycle $b \Rightarrow c$, $c \Rightarrow b$ occurs. Are there counterexamples with no cycles? Yes, but not for finite B, as we now show[1].

Define a replacement system $(B,\Rightarrow)$ to be acyclic if there are no $b, c \in B$, $b \neq c$, such that $b \Rightarrow^* c$, $c \Rightarrow^* b$. Then

2.1 *If* $(B,\Rightarrow)$ *is acyclic and* B *is finite then* (*) *is a necessary and sufficient condition for* $(B,\Rightarrow)$ *to be* CR.

To prove sufficiency consider arbitrary $a \in B$ and construct a tree with labelled nodes as follows:

(i) The base node is labelled a.

(ii) Immediately above a node labelled x is a node labelled y

---

[1] The results of 2.1 to 2.4 are due to M.H.A. Newman, Annals of Mathematics 43 (1942), 223-243.

Figure 1

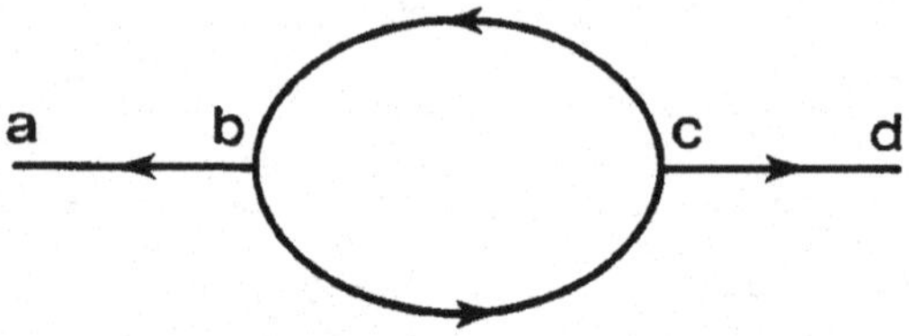

Figure 2

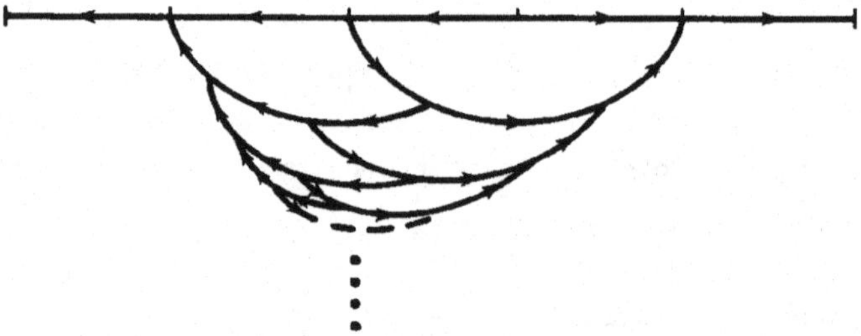

for each $y \in B$ such that $x \Rightarrow y$.

Observe that since $(B,\Rightarrow)$ is acyclic no two nodes on a branch of the tree have the same label, hence as B is finite every branch has finite length. Hence for every node of the tree there is a branch through that node ending in a terminal node. We conclude by showing that all terminal nodes have the same label. The proof is by induction on the number m of nodes in the longest branch of the tree. The case $m = 1$ is trivial, so we consider the case $m > 1$. If there is one node only immediately above the base node of the tree the inductive hypothesis applies immediately. Otherwise we may as well suppose that there are just two such nodes, say n_1 and n_2 labelled b and c respectively.

By the hypothesis (*) there is a node n_3 above n_1, and a node n_4 above n_2, which are both labelled d. Hence the subtrees above n_3 and n_4 are identical, in particular their terminal nodes are identically labelled, as required to complete the proof.

2.2 The restriction to finite systems can be weakened (see below) but not omitted as the counterexample in figure 2 shows. As in figure 1 the nodes represent elements of the system and the binary relation is indicated by directed arcs between nodes. The system of figure 2, as well as satisfying (*), also has the properties that every element has a normal form, and has only finitely many arcs leading to it and from it, yet it is clearly not CR.

2.3 Given an arbitrary replacement system $\mathcal{B}$ we can define as follows an acyclic system $\mathcal{B}_\alpha$, called the ordering of $\mathcal{B}$, which is CR if and only if $\mathcal{B}$ is CR. Define $\mathcal{B}_\alpha = (B_\alpha, \Rightarrow_\alpha)$ where

(i) B_α is the set of equivalence classes of elements of B with respect to the following equivalence relation; a, b are equivalent if and only if $a \Rightarrow^* b$ and $b \Rightarrow^* a$.

(ii) $a \Rightarrow_\alpha b$ if and only if there are $a' \in a$, $b' \in b$ such that $a' \Rightarrow b'$.

A replacement system is called locally finite if for all $a \in B$ the set $B_a = \{b : a \Rightarrow^* b\}$ is finite. Write $\Rightarrow_a$ for the restriction of $\Rightarrow$ to B_a; the system $(B_a, \Rightarrow_a)$ is called the local system below a. It now follows easily from the above remarks that

2.4 *A replacement system with locally finite ordering is* CR *if and only if its ordering satisfies* (*).

The result 2.1 can also be extended in another direction as follows.

2.5 *Suppose that* $(B,\Rightarrow)$ *is a replacement system such that there is a function* $b \mapsto n_b$ *from* B *to the natural numbers such that for all* $b,c \in B$, $b \Rightarrow c$ *implies* $n_b \geq n_c$ *and*

(i) $n_b > 0$ *implies* $n_b > n_c$

(ii) $n_b = 0$ *implies that the local system below* b *is* CR.

Then $(B,\Rightarrow)$ *is* CR *if it satisfies* (*).

The proof consists in showing that the local system below each $a \in B$ is CR, by induction on n_a. The case $n_a = 0$ is given by hypothesis so suppose that $n_a > 0$.

To prove that the local system below a is CR it is by inductive hypothesis sufficient to prove for arbitrary $a \Rightarrow^* b$, $a \Rightarrow^* c$ that there is d such that $b \Rightarrow^* d$, $c \Rightarrow^* d$, since for $a' \neq a$ such that $a \Rightarrow^* a'$ $n_{a'} < n_a$ so the local system below a' is CR by inductive hypothesis. The cases $a = b$ or $a = c$ are trivial, so we suppose that there are a_1, a_2 such that $a \Rightarrow a_1$, $a_1 \Rightarrow^* b$, $a \Rightarrow a_2$, $a_2 \Rightarrow^* c$. From (*) there is e such that $a_1 \Rightarrow^* e$, $a_2 \Rightarrow^* e$. Since the local systems below a_1, a_2 are CR there are f, g such that $b \Rightarrow^* f$, $e \Rightarrow^* f$, $c \Rightarrow^* g$, $e \Rightarrow^* g$. Hence as the local system below e is also CR there is d such that $f \Rightarrow^* d$, $g \Rightarrow^* d$, hence $b \Rightarrow^* d$, $c \Rightarrow^* d$ as required.

3. Refinements

A system $\mathcal{B}_2 = (B,\Rightarrow_2)$ is called a refinement of $\mathcal{B}_1 = (B,\Rightarrow_1)$ if $\Rightarrow_1 \subseteq \Rightarrow_2^*$. Such a refinement is called compatible if for all $a \Rightarrow_2^* b$ there is $c \in B$ such that $b \Rightarrow_1^* c$, $a \Rightarrow_1^* c$. Observe that

3.1 *A refinement* $(B,\Rightarrow_2)$ *of a system* $(B,\Rightarrow_1)$ *is compatible if and only if for all* $a \Rightarrow_2 b$, $b \Rightarrow_1^* c$ *there is* d *such that* $c \Rightarrow_1^* d$, $a \Rightarrow_1^* d$.

One part of the assertion is trivial since if $a \Rightarrow_2 b$, $b \Rightarrow_1^* c$ then $a \Rightarrow_2^* c$. We prove the converse. In particular we prove that if $a \Rightarrow_2 a_1 \Rightarrow_2 \dots \Rightarrow_2 a_K = b$ then there is $c \in B$ such that $b \Rightarrow_1^* c$, $a \Rightarrow_1^* c$; the argument is by induction on K. The case $K = 1$ is given by the hypothesis, and by inductive hypothesis there is $e \in B$ such that $b \Rightarrow_1^* e$, $a_1 \Rightarrow_1^* e$. Hence we can apply the hypothesis again to obtain that there is d such that $e \Rightarrow_1^* d$, $a \Rightarrow_1^* d$, so $a \Rightarrow_1^* d$, $b \Rightarrow_1^* d$ as required.

3.2 *If* $(B,\Rightarrow_2)$ *is a refinement of a* CR *system* $(B,\Rightarrow_1)$ *then* $(B,\Rightarrow_2)$ *is compatible if and only if for all* $a \Rightarrow_2 b$ *there is* $c \in B$ *such that* $a \Rightarrow_1^* c$, $b \Rightarrow_1^* c$.

One part of 3.2 is trivial. For the converse we prove by induction on K that if $a \Rightarrow_2 a_1 \Rightarrow_2 \ldots \Rightarrow_2 a_K = b$ then there is $c \in B$ such that $a \Rightarrow_1^* c$, $b \Rightarrow_1^* c$. The case $K = 1$ is given by hypothesis so we suppose that $K > 1$. By hypothesis there is $e \in B$ such that $a_1 \Rightarrow_1^* e$, $a \Rightarrow_1^* e$, and by inductive hypothesis there is $f \in B$ such that $a_1 \Rightarrow_1^* f$, $b \Rightarrow^* f$. Since $(B, \Rightarrow_1)$ is CR we therefore conclude that there is $c \in B$ such that $e \Rightarrow_1^* c$, $f \Rightarrow_1^* c$, so that $a \Rightarrow_1^* c$, $b \Rightarrow_1^* c$ as required.

3.3 Compatible refinements are interesting for the following reason.

A replacement system $(B,\Rightarrow_1)$ *is* CR *if and only if it has a compatible refinement which is* CR *and in that case every compatible refinement of* $(B,\Rightarrow_1)$ *is* CR.

Every system is a compatible refinement of itself so to prove this assertion it is sufficient to consider a particular compatible refinement $(B, \Rightarrow_2)$ of $(B, \Rightarrow_1)$ and to prove that the former is CR if and only if the latter is.

Suppose first that $(B, \Rightarrow_2)$ is CR and consider $a \Rightarrow_1^* b$, $a \Rightarrow_1^* c$; since $\Rightarrow_1^* \subseteq \Rightarrow_2^*$ there is, by hypothesis, $d \in B$ such that $b \Rightarrow_2^* d$, $c \Rightarrow_2^* d$. As $(B,\Rightarrow_2)$ is a compatible refinement of $(B, \Rightarrow_1)$ there is $e \in B$ such that $d \Rightarrow_1^* e$, $b \Rightarrow_1^* e$. As $\Rightarrow_1^* \subseteq \Rightarrow_2^*$ then $c \Rightarrow_2^* e$ so similarly there is $f \in B$ such that $e \Rightarrow_1^* f$, $c \Rightarrow_1^* f$. Hence $b \Rightarrow_1^* f$, $c \Rightarrow_1^* f$ as required.

Now suppose conversely that $(B,\Rightarrow_1)$ is CR. It is sufficient from 1.4 to find, given $a \Rightarrow_2^* b$ and $a \Rightarrow_2 c$, $d \in B$ such that $b \Rightarrow_2^* d$, $c \Rightarrow_2^* d$. We suppose in particular that $a \Rightarrow_2 a_1 \Rightarrow_2 \dots \Rightarrow_2 a_K=b$ and argue by induction on K, the case $K = 0$ (that is, $a = b$) being trivial.

Suppose then that $K = L + 1$. By inductive hypothesis there is $e \in B$ such that $a_L \Rightarrow_2^* e$, $c \Rightarrow_2^* e$. By hypothesis there are f, $g \in B$ such that $a_L \Rightarrow_1^* f$, $b \Rightarrow_1^* f$, $e \Rightarrow_1^* g$, $a_L \Rightarrow_1^* g$. Since $(B,\Rightarrow_1)$ is CR there is therefore $d \in B$ such that $f \Rightarrow_1^* d$, $g \Rightarrow_1^* d$ so as $\Rightarrow_1^* \subseteq \Rightarrow_2^*$ we have $b \Rightarrow_2^* d$, $c \Rightarrow_2^* d$ as required.

4. *Generalizations of the Mitschke-Rosen method*

Mitschke [73] and Rosen [73] independently gave a method for proving the lambda calculus with beta reduction to be CR, and this method has been analyzed by Barendregt [71]. He used it to give a new proof of the CR theorem for weak combinatory logic. Barendregt's analysis suggests the following generalizations of the method.

4.1 Suppose that $\mathcal{B}_j = (B_j, \Rightarrow_j)$, $j = 1,2$ are replacement systems and $i : B_1 \to B_2$, $\theta: B_2 \to B_1$ are maps. Write, for $a,b \in B_1$, $a \Rightarrow^0 b$ if there is $c \in B_2$ such that $i(a) \Rightarrow_2^* c$ and $\theta(c) = b$.

If the following conditions are satisfied and $\mathcal{B}_2$ *is* CR *then* $\mathcal{B}_1$ *is* CR:

(i) $a \Rightarrow_1 b$ *implies* $a \Rightarrow^0 b$

(ii) $a \Rightarrow_2 b$ *implies* $\theta(a) \Rightarrow_1^* \theta(b)$

(iii) $a \Rightarrow_2^* b$ *implies there is* c *such that* $\theta(a) \Rightarrow^0 \theta(c)$ *and* $\theta(b) \Rightarrow_1^* \theta(c)$.

To prove this it is sufficient in view of 1.4 and (i) to prove that if $a \Rightarrow^0 b$, $a \Rightarrow_1^* c$ then there is $d \in B_1$ such that $b \Rightarrow_1^* d$, $c \Rightarrow_1^* d$. The latter may be proved inductively, in view of (ii) and (iii), if one proves that for all $a \Rightarrow^0 b$, $a \Rightarrow_1 c$ there are $e, f \in B_2$ such that $\theta(e) = c$, $e \Rightarrow_2^* f$ and $b \Rightarrow_1^* \theta(f)$.

Hence suppose that $a \Rightarrow^0 b$, $a \Rightarrow_1 c$. By hypothesis and (i) there are $g, e \in B_2$ such that $i(a) \Rightarrow_2^* g$, $b = \theta(g)$, $i(a) \Rightarrow_2^* e$, $\theta(e) = c$. Since by hypothesis $\Rightarrow_2$ is CR there is $f \in B_2$ such that $g \Rightarrow_2^* f$, $e \Rightarrow_2^* f$ and from (ii) $b \Rightarrow_1^* \theta(f)$ as required.

4.2 Note that in 4.1 the hypothesis that $\Rightarrow_2$ be CR can be weakened; for the above proof we need only that for all $a \in B_1$ and $i(a) \Rightarrow_2^* b$, $i(a) \Rightarrow_2^* c$ there are $d, e \in B_2$ such that $b \Rightarrow_2^* d$, $c \Rightarrow_2^* e$ and $\theta(d) = \theta(e)$.

4.3 A slightly different generalization, in which θ is not required to be single-valued, turns out to be also a generalization of Rosen's condition 1.5 for the union of two CR systems to be CR.

Suppose that $\mathcal{B}_j = (B_j, \Rightarrow_j)$, $j = 1,2$ are replacement systems, $i : B_1 \to B_2$ is a map and θ is a map from B_2 into the subsets of B_1. Write $a \Rightarrow^0 b$ to mean that there is $c \in B_2$ such that $i(a) \Rightarrow_2^* c$ and $b \in \theta(c)$.

If the following conditions are satisfied and $\Rightarrow_2$ is CR then $\Rightarrow_1$ is CR; note that the conditions require that in general B_1 and B_2 overlap.

(i) $a \Rightarrow_1 b$ implies $a \Rightarrow^0 b$

(ii) $a \Rightarrow_2 b$ and $c \in \theta(a)$ implies there is $d \in \theta(b)$ such that $c \Rightarrow_1^* d$

(iii) if $a \Rightarrow_2^* b$ and $c \in \theta(a)$ then there is $d \in \theta(b)$ and $e \in B_2$ such that $i(c) \Rightarrow_2^* e$ and $d \in \theta(e)$

(iv) if $b \in \theta(a)$, $c \in \theta(a)$ and $b \neq c$ then $\theta(a) \subseteq B_2$ and there is $d \in \theta(b) \cap \theta(c)$

(v) if $b \in \theta(a)$, $b \in B_2$ and $c \in \theta(b)$, then $c \in \theta(a)$.

(vi) if $a \Rightarrow_1^* b$, $b \in B_2$ and $c \in \theta(b)$ then $a \Rightarrow_1^* c$.

As for the proof of 4.1 it is enough to prove that if $a \Rightarrow^0 b$, $a \Rightarrow_1 c$ then there is $h \in B_1$ such that $b \Rightarrow_1^* h$, $c \Rightarrow^0 h$, and $k \in B_1$ such that $b \Rightarrow_1^* k$, $c \Rightarrow_1^* k$.

There are $g,e \in B_2$ such that $i(a) \Rightarrow_2^* g$, $b \in \theta(g)$, $i(a) \Rightarrow_2^* e$, $c \in \theta(e)$. Since $\Rightarrow_2$ is CR there is $f \in B_2$ such that $g \Rightarrow_2^* f$, $e \Rightarrow_2^* f$. From (ii) and (iii) there are $m,n \in \theta(f)$ such that $b \Rightarrow_1^* m$, $c \Rightarrow^0 n$. If $m = n$, put $h = m$. If $m \neq n$, from (iv) there is $h \in \theta(m) \cap \theta(n)$. From (vi) $b \Rightarrow_1^* h$, and from (v) $c \Rightarrow^0 h$. Similarly there is $n' \in \theta(f)$ such that $c \Rightarrow_1^* n'$, hence there is similarly $k \in \theta(f)$ such that $b \Rightarrow_1^* k$, $c \Rightarrow_1^* k$ as required.

4.4 Consider the following application of 4.3. Suppose that $\mathcal{B}_j = (B,\Rightarrow_j)$, $j = p,q$ are two replacement systems and define $i : B \to B$ to be the identity map and define θ by $b \in \theta(a)$ if and only if $a \Rightarrow_q^* b$. We apply 4.3 by taking $B_1 = B_2 = b$, $\Rightarrow_1$ to be the union of the binary relations $\Rightarrow_p$ and $\Rightarrow_q$, and $\Rightarrow_2$ to be $\Rightarrow_p$. Clearly conditions (i), (v) and (vi) of 4.3 are satisfied. Observe that in the terminology of 1.6,

(ii) *and* (iii) *are satisfied if and only if* $(B,\Rightarrow_p)$ *requests* $(B, \Rightarrow_q)$.

Indeed in our notation $(B,\Rightarrow_p)$ requests $(B,\Rightarrow_q)$ means that if $a \Rightarrow_p^* b$ and $c \in \theta(a)$ then there is $d \in \theta(b)$ and $e \in B$ such that $c \Rightarrow_p^* e$, $d \in \theta(e)$. Since this is just (iii), one part is trivial.

If conversely $(B,\Rightarrow_p)$ requests $(B,\Rightarrow_q)$ then (iii) is trivially satisfied, and (ii) is also since from the definition of $\Rightarrow_1$, (ii) is a special case of (iii). Hence the stated result.

It follows that 4.3 is a generalization of 1.6. Note however that Rosen's sufficient condition 1.7 for one system to request another apparently has no analogous generalization.

4.5 As in 4.2 we can weaken in 4.3 the condition on $(B_2,\Rightarrow_2)$ that it be CR. It is sufficient to require that if $a \in B_1$, $i(a) \Rightarrow_2^* b$, $i(a) \Rightarrow_2^* c$ then there are d, e such that $b \Rightarrow_2^* d$, $c \Rightarrow_2^* e$ and $\theta(d) \cap \theta(e)$ is nonempty. This generalizes the weakening of 1.6 given in 1.8.

4.6 There are further, slightly different ways to generalize the Mitschke-Rosen method; for example the following.

Suppose that $\mathcal{B}_j = (B_j,\Rightarrow_j)$, $j = 1,2$ are replacement systems and $i : B_1 \to B_2$, $\theta : B_2 \to B_1$ are maps. Suppose also that

(i) $a \Rightarrow_1 b$ implies there is c such that $i(a) \Rightarrow_2^* c$ and $\theta(c) = b$

(ii) $i(a) \Rightarrow_2^* c$ implies $a \Rightarrow_1^* \theta(c)$

(iii) for all $a \in B_1$ there is $d_a \in B_2$ such that $i(a) \Rightarrow_2^* b$, $i(a) \Rightarrow_2^* c$ implies $b \Rightarrow_2^* d_a$, $c \Rightarrow_2^* d_a$, and

(iv) $i(a) \Rightarrow_2^* b \Rightarrow_2^* d_a$ implies there is e such that $i(\theta(b)) \Rightarrow_2^* e$ and $\theta(e) = \theta(d_a)$;

then $(B_1,\Rightarrow_1)$ is CR.

The proof differs only slightly from the proof of 4.1 and so it is omitted.

5. *Open problems arising from a question of Mann*

Even for the replacement system of classical interest open CR problems arise. Roger Hindley has mentioned in correspondence the following question, originally asked by Colin Mann; is the extension of the lambda calculus which is defined in 5.1 CR? One can ask the same question (5.2) for weak combinatory logic, for example. Hindley suggested that a simpler question (5.3) might be simpler to answer; of course that question can also be asked for weak combinatory logic (5.4). An even simpler extension, still with unresolved CR problem, is given in 5.5 and 5.6.

The lambda calculus concerned is the usual λ-K-calculus of Curry and Feys [58], with β-reduction. Weak combinatory logic is based for definiteness on the two primitive combinators S and K, with the rule schemes $SXYZ \to XZ(YZ)$ and $KXY \to X$.

5.1 Add to the lambda calculus new primitive symbols D, D_1 and D_2 and add as new rules those which satisfy the schemes

$D(D_1X)(D_2X) \to X$

$D_1(DXY) \to X$

$D_2(DXY) \to Y$. Is the new calculus CR?

5.2 Extend weak combinatory logic in the same way as 5.1. Is it CR?

5.3 Add to the lambda calculus a new primitive symbol D and the rules given by the scheme

DXX → X. Is the extended calculus CR?

5.4 Is the extension of weak combinatory logic corresponding to 5.3 CR?

5.5 Add to the lambda calculus two new primitive symbols D and E and the rules given by the scheme

DXX → E. Is the extended calculus CR?

5.6 Is the extension of weak combinatory logic corresponding to 5.5 CR?

6. *Acknowledgement*

I thank Roger Hindley for telling me about the calculi of 5.1 and 5.3, and for helpful criticism during the preparation of this paper.

BIBLIOGRAPHY

BARENDREGT, H.P.

[71] Some extensional term models for combinatory logics and λ-calculi. Ph.D. thesis, U. Utrecht, 1971.

CURRY, H.B. and R. FEYS

[58] Combinatory logic, North-Holland, Amsterdam, 1958.

HINDLEY, R.

[64] The Church-Rosser property and a result in combinatory logic. Ph.D. thesis, U. Newcastle-upon-Tyne, 1964.

[69] An abstract form of the Church-Rosser theorem, I. J. Symbolic Logic 34, 1969 , 545-560.

[74] An abstract Church-Rosser theorem, II: applications. J. Symbolic Logic, 39, 1974, 1-21.

MITSCHKE, G.

[73] Ein algebraischer Beweis für das Church-Rosser Theorem, Arch. math. Logik 15, 1973, 146-157.

ROSEN, B.K.

[73] Tree-manipulating systems and Church-Rosser theorems. J.A.C.M. 20, 1973, 160-187.

SCHROER, D.E.

[65] The Church-Rosser theorem. Ph.D. thesis, Cornell University, 1965.

* * *

Department of Mathematics, Australian National University, Canberra, Australia.

Vol. 277: Séminaire Banach. Edité par C. Houzel. VII, 229 pages. 1972. DM 22,-

Vol. 278: H. Jacquet, Automorphic Forms on GL(2) Part II. XIII, 142 pages. 1972. DM 18,-

Vol. 279: R. Bott, S. Gitler and I. M. James, Lectures on Algebraic and Differential Topology. V, 174 pages. 1972. DM 20,-

Vol. 280: Conference on the Theory of Ordinary and Partial Differential Equations. Edited by W. N. Everitt and B. D. Sleeman. XV, 367 pages. 1972. DM 29,-

Vol. 281: Coherence in Categories. Edited by S. Mac Lane. VII, 235 pages. 1972. DM 22,-

Vol. 282: W. Klingenberg und P. Flaschel, Riemannsche Hilbertmannigfaltigkeiten. Periodische Geodätische. VII, 211 Seiten. 1972. DM 22,-

Vol. 283: L. Illusie, Complexe Cotangent et Déformations II. VII, 304 pages. 1972. DM 27,-

Vol. 284: P. A. Meyer, Martingales and Stochastic Integrals I. VI, 89 pages. 1972. DM 18,-

Vol. 285: P. de la Harpe, Classical Banach-Lie Algebras and Banach-Lie Groups of Operators in Hilbert Space. III, 160 pages. 1972. DM 18,-

Vol. 286: S. Murakami, On Automorphisms of Siegel Domains. V, 95 pages. 1972. DM 18,-

Vol. 287: Hyperfunctions and Pseudo-Differential Equations. Edited by H. Komatsu. VII, 529 pages. 1973. DM 40,-

Vol. 288: Groupes de Monodromie en Géométrie Algébrique. (SGA 7 I). Dirigé par A. Grothendieck. IX, 523 pages. 1972. DM 55,-

Vol. 289: B. Fuglede, Finely Harmonic Functions. III, 188. 1972. DM 20,-

Vol. 290: D. B. Zagier, Equivariant Pontrjagin Classes and Applications to Orbit Spaces. IX, 130 pages. 1972. DM 18,-

Vol. 291: P. Orlik, Seifert Manifolds. VIII, 155 pages. 1972. DM 18,-

Vol. 292: W. D. Wallis, A. P. Street and J. S. Wallis, Combinatorics: Room Squares, Sum-Free Sets, Hadamard Matrices. V, 508 pages. 1972. DM 55,-

Vol. 293: R. A. DeVore, The Approximation of Continuous Functions by Positive Linear Operators. VIII, 289 pages. 1972. DM 27,-

Vol. 294: Stability of Stochastic Dynamical Systems. Edited by R. F. Curtain. IX, 332 pages. 1972. DM 29,-

Vol. 295: C. Dellacherie, Ensembles Analytiques Capacités. Mesures de Hausdorff. XII, 123 pages. 1972. DM 18,-

Vol. 296: Probability and Information Theory II. Edited by M. Behara, K. Krickeberg and J. Wolfowitz. V, 223 pages. 1973. DM 22,-

Vol. 297: J. Garnett, Analytic Capacity and Measure. IV, 138 pages. 1972. DM 18,-

Vol. 298: Proceedings of the Second Conference on Compact Transformation Groups. Part 1. XIII, 453 pages. 1972. DM 35,-

Vol. 299: Proceedings of the Second Conference on Compact Transformation Groups. Part 2. XIV, 327 pages. 1972. DM 29,-

Vol. 300: P. Eymard, Moyennes Invariantes et Représentations Unitaires. II, 113 pages. 1972. DM 18,-

Vol. 301: F. Pittnauer, Vorlesungen über asymptotische Reihen. VI, 186 Seiten. 1972. DM 18,-

Vol. 302: M. Demazure, Lectures on p-Divisible Groups. V, 98 pages. 1972. DM 18,-

Vol. 303: Graph Theory and Applications. Edited by Y. Alavi, D. R. Lick and A. T. White. IX, 329 pages. 1972. DM 26,-

Vol. 304: A. K. Bousfield and D. M. Kan, Homotopy Limits, Completions and Localizations. V, 348 pages. 1972. DM 29,-

Vol. 305: Théorie des Topos et Cohomologie Etale des Schémas. Tome 3. (SGA 4). Dirigé par M. Artin, A. Grothendieck et J. L. Verdier. VI, 640 pages. 1973. DM 55,-

Vol. 306: H. Luckhardt, Extensional Gödel Functional Interpretation. VI, 161 pages. 1973. DM 20,-

Vol. 307: J. L. Bretagnolle, S. D. Chatterji et P.-A. Meyer, Ecole d'été de Probabilités: Processus Stochastiques. VI, 198 pages. 1973. DM 22,-

Vol. 308: D. Knutson, λ-Rings and the Representation Theory of the Symmetric Group. IV, 203 pages. 1973. DM 22,-

Vol. 309: D. H. Sattinger, Topics in Stability and Bifurcation Theory. VI, 190 pages. 1973. DM 20,-

Vol. 310: B. Iversen, Generic Local Structure of the Morphisms in Commutative Algebra. IV, 108 pages. 1973. DM 18,-

Vol. 311: Conference on Commutative Algebra. Edited by J. W. Brewer and E. A. Rutter. VII, 251 pages. 1973. DM 24,-

Vol. 312: Symposium on Ordinary Differential Equations. Edited by W. A. Harris, Jr. and Y. Sibuya. VIII, 204 pages. 1973. DM 22,-

Vol. 313: K. Jörgens and J. Weidmann, Spectral Properties of Hamiltonian Operators. III, 140 pages. 1973. DM 18,-

Vol. 314: M. Deuring, Lectures on the Theory of Algebraic Functions of One Variable. VI, 151 pages. 1973. DM 18,-

Vol. 315: K. Bichteler, Integration Theory (with Special Attention to Vector Measures). VI, 357 pages. 1973. DM 29,-

Vol. 316: Symposium on Non-Well-Posed Problems and Logarithmic Convexity. Edited by R. J. Knops. V, 176 pages. 1973. DM 20,-

Vol. 317: Séminaire Bourbaki - vol. 1971/72. Exposés 400-417. IV, 361 pages. 1973. DM 29,-

Vol. 318: Recent Advances in Topological Dynamics. Edited by A. Beck. VIII, 285 pages. 1973. DM 27,-

Vol. 319: Conference on Group Theory. Edited by R. W. Gatterdam and K. W. Weston. V, 188 pages. 1973. DM 20,-

Vol. 320: Modular Functions of One Variable I. Edited by W. Kuyk. V, 195 pages. 1973. DM 20,-

Vol. 321: Séminaire de Probabilités VII. Edité par P. A. Meyer. VI, 322 pages. 1973. DM 29,-

Vol. 322: Nonlinear Problems in the Physical Sciences and Biology. Edited by I. Stakgold, D. D. Joseph and D. H. Sattinger. VIII, 357 pages. 1973. DM 29,-

Vol. 323: J. L. Lions, Perturbations Singulières dans les Problèmes aux Limites et en Contrôle Optimal. XII, 645 pages. 1973. DM 46,-

Vol. 324: K. Kreith, Oscillation Theory. VI, 109 pages. 1973. DM 18,-

Vol. 325: C.-C. Chou, La Transformation de Fourier Complexe et L'Equation de Convolution. IX, 137 pages. 1973. DM 18,-

Vol. 326: A. Robert, Elliptic Curves. VIII, 264 pages. 1973. DM 24,-

Vol. 327: E. Matlis, One-Dimensional Cohen-Macaulay Rings. XII, 157 pages. 1973. DM 20,-

Vol. 328: J. R. Büchi and D. Siefkes, The Monadic Second Order Theory of All Countable Ordinals. VI, 217 pages. 1973. DM 22,-

Vol. 329: W. Trebels, Multipliers for (C, α)-Bounded Fourier Expansions in Banach Spaces and Approximation Theory. VII, 103 pages. 1973. DM 18,-

Vol. 330: Proceedings of the Second Japan-USSR Symposium on Probability Theory. Edited by G. Maruyama and Yu. V. Prokhorov. VI, 550 pages. 1973. DM 40,-

Vol. 331: Summer School on Topological Vector Spaces. Edited by L. Waelbroeck. VI, 226 pages. 1973. DM 22,-

Vol. 332: Séminaire Pierre Lelong (Analyse) Année 1971-1972. V, 131 pages. 1973. DM 18,-

Vol. 333: Numerische, insbesondere approximationstheoretische Behandlung von Funktionalgleichungen. Herausgegeben von R. Ansorge und W. Törnig. VI, 296 Seiten. 1973. DM 27,-

Vol. 334: F. Schweiger, The Metrical Theory of Jacobi-Perron Algorithm. V, 111 pages. 1973. DM 18,-

Vol. 335: H. Huck, R. Roitzsch, U. Simon, W. Vortisch, R. Walden, B. Wegner und W. Wendland, Beweismethoden der Differentialgeometrie im Großen. IX, 159 Seiten. 1973. DM 20,-

Vol. 336: L'Analyse Harmonique dans le Domaine Complexe. Edité par E. J. Akutowicz. VIII, 169 pages. 1973. DM 20,-

Vol. 337: Cambridge Summer School in Mathematical Logic. Edited by A. R. D. Mathias and H. Rogers. IX, 660 pages. 1973. DM 46,-

Vol. 338: J. Lindenstrauss and L. Tzafriri, Classical Banach Spaces. IX, 243 pages. 1973. DM 24,-

Vol. 339: G. Kempf, F. Knudsen, D. Mumford and B. Saint-Donat, Toroidal Embeddings I. VIII, 209 pages. 1973. DM 22,-

Vol. 340: Groupes de Monodromie en Géométrie Algébrique. (SGA 7 II). Par P. Deligne et N. Katz. X, 438 pages. 1973. DM 44,-

Vol. 341: Algebraic K-Theory I, Higher K-Theories. Edited by H. Bass. XV, 335 pages. 1973. DM 29,-

Vol. 342: Algebraic K-Theory II, "Classical" Algebraic K-Theory, and Connections with Arithmetic. Edited by H. Bass. XV, 527 pages. 1973. DM 40,-

Vol. 343: Algebraic K-Theory III, Hermitian K-Theory and Geometric Applications. Edited by H. Bass. XV, 572 pages. 1973. DM 40,-

Vol. 344: A. S. Troelstra (Editor), Metamathematical Investigation of Intuitionistic Arithmetic and Analysis. XVII, 485 pages. 1973. DM 38,-

Vol. 345: Proceedings of a Conference on Operator Theory. Edited by P. A. Fillmore. VI, 228 pages. 1973. DM 22,-

Vol. 346: Fučík et al., Spectral Analysis of Nonlinear Operators. II, 287 pages. 1973. DM 26,-

Vol. 347: J. M. Boardman and R. M. Vogt, Homotopy Invariant Algebraic Structures on Topological Spaces. X, 257 pages. 1973. DM 24,-

Vol. 348: A. M. Mathai and R. K. Saxena, Generalized Hypergeometric Functions with Applications in Statistics and Physical Sciences. VII, 314 pages. 1973. DM 26,-

Vol. 349: Modular Functions of One Variable II. Edited by W. Kuyk and P. Deligne. V, 598 pages. 1973. DM 38,-

Vol. 350: Modular Functions of One Variable III. Edited by W. Kuyk and J.-P. Serre. V, 350 pages. 1973. DM 26,-

Vol. 351: H. Tachikawa, Quasi-Frobenius Rings and Generalizations. XI, 172 pages. 1973. DM 20,-

Vol. 352: J. D. Fay, Theta Functions on Riemann Surfaces. V, 137 pages. 1973. DM 18,-

Vol. 353: Proceedings of the Conference on Orders, Group Rings and Related Topics. Organized by J. S. Hsia, M. L. Madan and T. G. Ralley. X, 224 pages. 1973. DM 22,-

Vol. 354: K. J. Devlin, Aspects of Constructibility. XII, 240 pages. 1973. DM 24,-

Vol. 355: M. Sion, A Theory of Semigroup Valued Measures. V, 140 pages. 1973. DM 18,-

Vol. 356: W. L. J. van der Kallen, Infinitesimally Central Extensions of Chevalley Groups. VII, 147 pages. 1973. DM 18,-

Vol. 357: W. Borho, P. Gabriel und R. Rentschler, Primideale in Einhüllenden auflösbarer Lie-Algebren. V, 182 Seiten. 1973. DM 20,-

Vol. 358: F. L. Williams, Tensor Products of Principal Series Representations. VI, 132 pages. 1973. DM 18,-

Vol. 359: U. Stammbach, Homology in Group Theory. VIII, 183 pages. 1973. DM 20,-

Vol. 360: W. J. Padgett and R. L. Taylor, Laws of Large Numbers for Normed Linear Spaces and Certain Fréchet Spaces. VI, 111 pages. 1973. DM 18,-

Vol. 361: J. W. Schutz, Foundations of Special Relativity: Kinematic Axioms for Minkowski Space Time. XX, 314 pages. 1973. DM 26,-

Vol. 362: Proceedings of the Conference on Numerical Solution of Ordinary Differential Equations. Edited by D. Bettis. VIII, 490 pages. 1974. DM 34,-

Vol. 363: Conference on the Numerical Solution of Differential Equations. Edited by G. A. Watson. IX, 221 pages. 1974. DM 20,-

Vol. 364: Proceedings on Infinite Dimensional Holomorphy. Edited by T. L. Hayden and T. J. Suffridge. VII, 212 pages. 1974. DM 20,-

Vol. 365: R. P. Gilbert, Constructive Methods for Elliptic Equations. VII, 397 pages. 1974. DM 26,-

Vol. 366: R. Steinberg, Conjugacy Classes in Algebraic Groups (Notes by V. V. Deodhar). VI, 159 pages. 1974. DM 18,-

Vol. 367: K. Langmann und W. Lütkebohmert, Cousinverteilungen und Fortsetzungssätze. VI, 151 Seiten. 1974. DM 16,-

Vol. 368: R. J. Milgram, Unstable Homotopy from the Stable Point of View. V, 109 pages. 1974. DM 16,-

Vol. 369: Victoria Symposium on Nonstandard Analysis. Edited by A. Hurd and P. Loeb. XVIII, 339 pages. 1974. DM 26,-

Vol. 370: B. Mazur and W. Messing, Universal Extensions and One Dimensional Crystalline Cohomology. VII, 134 pages. 1974. DM 16,-

Vol. 371: V. Poenaru, Analyse Différentielle. V, 228 pages. 1974. DM 20,-

Vol. 372: Proceedings of the Second International Conference on the Theory of Groups 1973. Edited by M. F. Newman. VII, 740 pages. 1974. DM 48,-

Vol. 373: A. E. R. Woodcock and T. Poston, A Geometrical Study of the Elementary Catastrophes. V, 257 pages. 1974. DM 22,-

Vol. 374: S. Yamamuro, Differential Calculus in Topological Linear Spaces. IV, 179 pages. 1974. DM 18,-

Vol. 375: Topology Conference 1973. Edited by R. F. Dickman Jr. and P. Fletcher. X, 283 pages. 1974. DM 24,-

Vol. 376: D. B. Osteyee and I. J. Good, Information, Weight of Evidence, the Singularity between Probability Measures and Signal Detection. XI, 156 pages. 1974. DM 16,-

Vol. 377: A. M. Fink, Almost Periodic Differential Equations. VIII, 336 pages. 1974. DM 26,-

Vol. 378: TOPO 72 – General Topology and its Applications. Proceedings 1972. Edited by R. Alò, R. W. Heath and J. Nagata. XIV, 651 pages. 1974. DM 50,-

Vol. 379: A. Badrikian et S. Chevet, Mesures Cylindriques, Espaces de Wiener et Fonctions Aléatoires Gaussiennes. X, 383 pages. 1974. DM 32,-

Vol. 380: M. Petrich, Rings and Semigroups. VIII, 182 pages. 1974. DM 18,-

Vol. 381: Séminaire de Probabilités VIII. Edité par P. A. Meyer. IX, 354 pagès. 1974. DM 32,-

Vol. 382: J. H. van Lint, Combinatorial Theory Seminar Eindhoven University of Technology. VI, 131 pages. 1974. DM 18,-

Vol. 383: Séminaire Bourbaki – vol. 1972/73. Exposés 418-435 IV, 334 pages. 1974. DM 30,-

Vol. 384: Functional Analysis and Applications, Proceedings 1972. Edited by L. Nachbin. V, 270 pages. 1974. DM 22,-

Vol. 385: J. Douglas Jr. and T. Dupont, Collocation Methods for Parabolic Equations in a Single Space Variable (Based on C^1-Piecewise-Polynomial Spaces). V, 147 pages. 1974. DM 16,-

Vol. 386: J. Tits, Buildings of Spherical Type and Finite BN-Pairs. IX, 299 pages. 1974. DM 24,-

Vol. 387: C. P. Bruter, Eléments de la Théorie des Matroïdes. V, 138 pages. 1974. DM 18,-

Vol. 388: R. L. Lipsman, Group Representations. X, 166 pages. 1974. DM 20,-

Vol. 389: M.-A. Knus et M. Ojanguren, Théorie de la Descente et Algèbres d' Azumaya. IV, 163 pages. 1974. DM 20,-

Vol. 390: P. A. Meyer, P. Priouret et F. Spitzer, Ecole d'Eté de Probabilités de Saint-Flour III – 1973. Edité par A. Badrikian et P.-L. Hennequin. VIII, 189 pages. 1974. DM 20,-

Vol. 391: J. Gray, Formal Category Theory: Adjointness for 2-Categories. XII, 282 pages. 1974. DM 24,-

Vol. 392: Géométrie Différentielle, Colloque, Santiago de Compostela, Espagne 1972. Edité par E. Vidal. VI, 225 pages. 1974. DM 20,-

Vol. 393: G. Wassermann, Stability of Unfoldings. IX, 164 pages. 1974. DM 20,-

Vol. 394: W. M. Patterson 3rd, Iterative Methods for the Solution of a Linear Operator Equation in Hilbert Space – A Survey. III, 183 pages. 1974. DM 20,-

Vol. 395: Numerische Behandlung nichtlinearer Integrodifferential- und Differentialgleichungen. Tagung 1973. Herausgegeben von R. Ansorge und W. Törnig. VII, 313 Seiten. 1974. DM 28,-

Vol. 396: K. H. Hofmann, M. Mislove and A. Stralka, The Pontryagin Duality of Compact O-Dimensional Semilattices and its Applications. XVI, 122 pages. 1974. DM 18,-

Vol. 397: T. Yamada, The Schur Subgroup of the Brauer Group. V, 159 pages. 1974. DM 18,-

Vol. 398: Théories de l'Information, Actes des Rencontres de Marseille-Luminy, 1973. Edité par J. Kampé de Fériet et C. Picard. XII, 201 pages. 1974. DM 23,-